BASICS OF BIOTECHNOLOGY

BASICS OF BIOTECHNOLOGY

SITAKANTA BEHERA

ANMOL PUBLICATIONS PVT. LTD.
NEW DELHI - 110 002 (INDIA)

ANMOL PUBLICATIONS PVT. LTD.

H.O.: 4374/4B, Ansari Road, Darya Ganj,
New Delhi-110 002 (India)
Ph.: 23278000, 23261597

B.O.: No. 1015, Ist Main Road, BSK IIIrd Stage
IIIrd Phase, IIIrd Block
Bangalore - 560 085 (India)
Visit us at: www.anmolpublications.com

Basics of Biotechnology

ISBN 978-81-261-3495-3

PRINTED IN INDIA

Printed at Mehra Offset Press, Delhi.

Contents

Preface

The complexity of the ethical dilemmas posed by biotechnology is such that we are still trying to identify what questions should be asked. The volume recognizes that the first step in making the types of ethical decisions which are increasingly demanded of us is searching for the right questions. The chapters vary greatly in both length and orientation.

The Human Genome Project is generating fundamental genetic information at a breathtaking rate; basic research findings are being applied in medicine, agriculture, and the environment; and a variety of new biotechnology products are moving into production. Behind each of these accomplishments are teams of scientists and technicians whose everyday work makes such achievements possible.

This will make it useable by a wide range of groups — from students in the classroom, to professionals such as scientists, medical doctors, business people, lawyers, and ethicists, as well as environmental and other socially active groups. What becomes obvious from the book here is that the issues arising from biotechnology are dramatic and will impact all of us and thus are in need of wide-ranging public debate. It is hoped that this book will contribute to that debate.

Author

Chapter 1

Introduction

DEFINING "BIOTECHNOLOGY"

Advances in biotechnology are featured on the news and inspire the formation of many commissions and government advisory committees and even more legal wrangles. Biotechnology, the use of biological processes to produce a product or process for human use and benefit, is often portrayed as a revolutionary, even dangerous, way of viewing and using nature and its elements. It is also an ancient activity. The making of beer, wine and cheese are essentially biotechnologies, though some may question the benefit of wine and beer and indeed others may even question the benefit of cheese. The domestication, breeding and selection of plants and animals for human use using conventional methods also are biotechnologies.

1. The use of living things to make products.
2. The application of a biological process to produce a useful product.
3. Any technological application that uses biological systems, living organisms, or derivatives thereof, to make or modify products or processes for specific uses.

The use of microorganisms (such as bacteria or yeasts) or biological substances (such as enzymes) to perform specific industrial or manufacturing processes. Applications include the production of certain drugs, synthetic hormones, and bulk foodstuffs, as well as the bioconversion of organic waste and the cleanup of oil spills.

The application of the principles of engineering and technology to the life sciences—bioengineering.

Cloning, genetic manipulation, cell fusion, and mutation.

Essentially, doing "more and faster" what we have known and done for centuries. It is almost as old as agriculture itself.

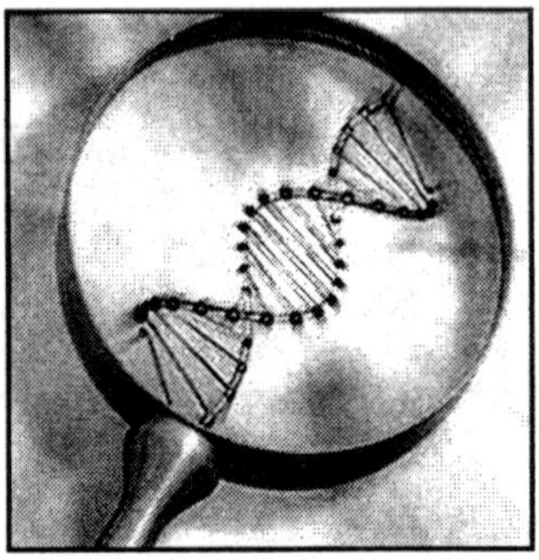

Fig. Magnification of Gene

Farmers and bakers are the biotechs of days gone by. Remember Grandma's freshly baked bread? How Grandpa kept the seeds of those really big pepper or tomatoes? Your grandparents were practicing biotechnology.

Modifying the genetic material of organisms directly and precisely. It enables the transfer of genes between diverse organisms, allowing combinations unlikely to occur by conventional means. Allowing speedier and more specific results.

Life- Define:

- Moving
- Growing
- Eating
- Breeding
- Response to stimuli

PRODUCTS

Traditional

1. Beer / Wine
2. Yeast
3. Ethanol
4. Cheese

Protein Therapeutics

1. Herceptin
2. Neupogen
3. Insulin - 1982 U.S. pharmaceutical manufacturer Eli Lilly the first to market genetically-engineered human insulin

Agriculture

1. Bacillus thuringiensis (Bt) Crops Bt crops
2. Herbicide tolerance
3. Disease resistance
4. Pest resistance

Industrial / Environmental

1. Hazardous Waste
2. Wastewater Treatment
3. Bioremediation

Other

1. Plastics
2. Biopulping
3. GMO (Genetically modified organism)

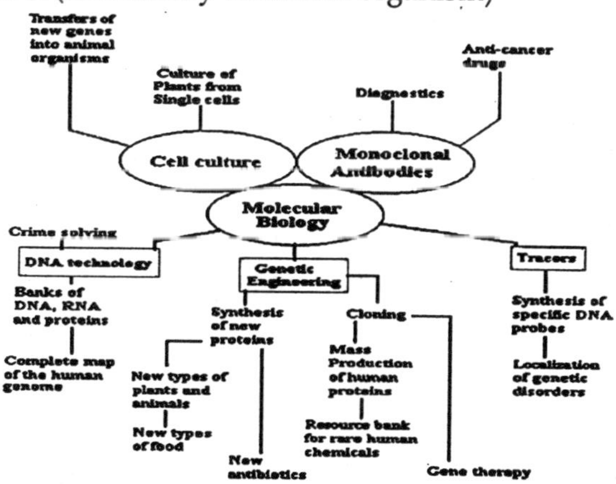

Fig. Concept of Biotechnology

The biotechnology activities that raise concerns and are the stuff of headlines and news programmes generally involve the application of recent dis-coveries in the biological sciences to alter the properties of a living organism by altering its genes, with the goal of producing a product or process for human use. Another term for the process of altering genes is genetic engineering, because the way in which useful biotechnologies are created often involves the alteration of the properties of an organism by engineering its genetic material.

The difference between science and engineering is important in thinking about biotechnology. Science can be defined as "the observation, identification, description, experi-mental investigation and theoretical explanation of natural phenomena". Engineering is "the application of scientific and mathe-matical principles to practical ends such as the design, manufacture and operation of efficient and economical structures, machines, processes and systems". Freeman Dyson, a physicist and scientist wrote, "A good scientist is a person with original ideas.

Fig. GMO

A good engineer is a person who makes a design that works with as few original ideas as possible". The key distinction is between understanding and using that understanding to make something. One event useful to mark the beginning of the modern, engineering-like and public era of biotechnology is the Asilomar conference on recombinant DNA in 1975.

Exploration of the history and stakeholders in the Asilomar Conference will provide ways to look at contemporary debates and dilemmas. However, to understand the issues surrounding the conference as well as the current controversies, some understanding of the basic concepts, language and tools of biotechnology and molecular biology are needed. This brief primer will provide that introduction.

Biotechnology could only arise as a practical undertaking when two elements were in place:

1. Sufficient understanding of the principles and processes that determined the properties of living organisms,
2. The tools to modify the properties of living organisms more or less at will—the science and the engineering. The scientific discoveries that provide the understanding of biological processes, the few good ideas of biotechnology, to draw from Dyson's line, can be traced back more than 150 years to include many discoveries and tools drawn from the disciplines of physics, chemistry and genetics and their intersections in biochemistry, bacteriology, physical chemistry and X-ray crystallography.

This prehistory of biotechnology particularly had an impact on two branches of biology: cell biology, the study of the structure and function of the simplest unit of life and genetics and the study of heredity, the mechanisms of transmission of traits from one generation to the next. The discoveries in these fields critical to biotechnology are those that eventually provided understanding at the molecular level of how the characteristics of an organism are inherited and expressed. The crucial discoveries, however, began at the level of the genetics of the whole organism. Modern genetics owes its beginnings to Gregor Mendel, an Austrian Augustinian monk trained in mathematics and botany, who in the 1860s performed experiments on the inheritance of the traits of pea plants.

A key to his success was that he limited his studies to categorical traits, traits with two and only two distinct forms. Mendel hypothesized that categorical traits of pea plants such as the colour of seeds (yellow or green), seed shape (smooth or wrinkled), length of the stem (tall or dwarf) were determined by what he called factors. One factor was inherited from each parent, and factors were expressed in the offspring in a dominant or recessive form.

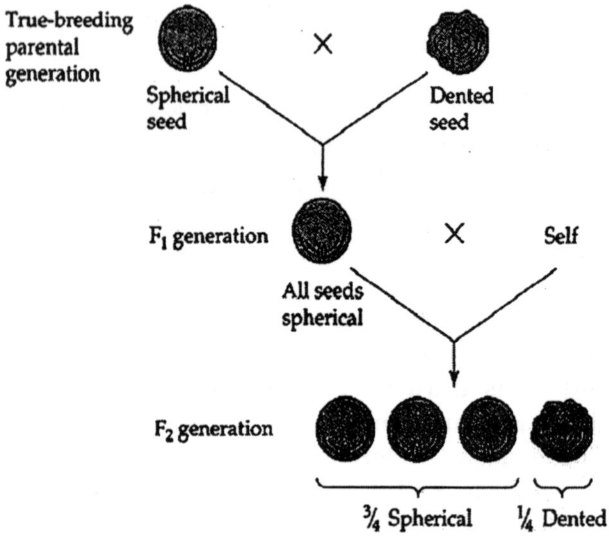

Fig. Mendel first experiment

Dominance requires only one copy of the factor in question to see the effect. Recessive factors required that both copies be the same to see the expression of the trait in the offspring. If an offspring inherited a dominant and a recessive factor, the plant would show the effect of the dominant factor, but could pass on the recessive factor to the next generation. He tested his hypothesis by crossing (mating) selected pea plants and counting the offspring in each generation that expressed the factor in question that showed the trait.

He was testing his idea by crossing particular plants, predicting the number of offspring of each type and comparing the results to the prediction. Mendel's work was precise and showed that the inheritance of categorical traits was predictable and subject to quantitative, statistical analyses. He carefully documented his experiments and published the results, but his report in the *Proceedings of the Society of Natural Sciences* remained essentially unnoticed until the turn of the century, when it was independently rediscovered by three botanists who repeated his work and obtained the same results.

Their recognition of his work and their publicizing of Mendel's insights provided the beginning of a rigorous science of genetics. The ensuing decades provided insights into the biology and then the biochemistry of cells and whole organisms. The research addresses three related questions:

1. How were the properties, the factors of Mendel, or genes in current language transmitted from generation to generation in a more or less faithful way?
2. What was the chemical nature of the genetic material?
3. How did the genetic material provide instruction for the properties of the cell, the tissue, the organism and the whole body?

While much remains unknown, particularly for question 3 for complex organisms such as humans, a useful understanding has been achieved for the first two questions.

Mendel had set the stage by deducing the rules of inheritance, but moving from a set of rules to an understanding of the chemical basis for the phenomena required knowledge of the structure of the cell and the chemistry of how it functioned. This effort was driven by the belief that the clearest understanding of how living organisms worked would come from an understanding at the simplest level, moving from the whole organism to an organ, a tissue, a cell and finally to the chemicals of which the cells are made.

This reductionist approach to biology was highly successful and paralleled discoveries in chemistry and physics, some of which provided not only tools, technologies and concepts but also some of the most creative scientific minds to fuel progress in biological research. Erwin Schrodinger, Max Delbruck, Linus Pauling, Max Perutz and Francis Crick were some of the physicists who applied their understanding of chemistry and physics, particularly thermodynamics, to properties of living organisms and used their insights to formulate new ways of thinking about living things.

In 1944 Schrodinger presented a series of lectures at Trinity College in Dublin, which were published as a short, but highly influential book entitled, *What Is Life?* He addressed two themes in biology, the nature of heredity and the thermodynamics of living systems. Many early researchers in what became molecular biology credit their interest in studying the molecular basis for heredity to the influence of Shrodinger's book. An important consequence of the reductionist approach to biology is that as understanding is expanded there is a presumption that the process may be amenable to the engineer's skills. What can be understood can be modified and used by humans for human benefit.

The Cell's Structure

The fundamental structural unit of all living organisms is the cell, the smallest unit capable of independent life. Viruses, which can only live by infecting cells, are an exception to this rule. Each cell is bounded by a membrane or cell wall. Cells come in many varieties. Each single-celled organism has particular properties. In multicelled organisms, from earthworms to humans, different cells play different roles and are specialized for that function. There are, however, recurring patterns of structure and function shared by all cells.

It is now recognized that there are three major forms or kingdoms of living organisms: archae, the most ancient form, able to grow under extreme conditions such as above 90°C; prokaryotes such as bacteria and certain algae; and eukaryotes, such as yeast, amoeba, plants and animals. The cells of eukaryotes have several important distinctions. Eukaryotes, but not prokaryotes or archae, have internal membrane-bound structures called organelles that carry out specific functions.

One of the organelles of eukaryotic cells is the nucleus that contains the structures carrying the genetic information, the chromosomes. In prokaryotes and archae the chromosome is not in a membrane-bound organelle but free within the cytoplasm. The mitochondrion, which functions to break down nutrients to provide raw materials and energy for cellular processes, is another organelle of eukaryotic cells.

While each eukaryotic cell has a single nucleus, there are several mitochondria in each cell. It is thought that in the evolution of the eukaryotic cell, a prokaryotic cell was taken into the cytoplasm, enclosed by both its own membrane and the membrane of the engulfing cell. The prokaryote's membrane was not broken down. Mitochon-dria have a double membrane and carry their own genetic information, supporting this theory. Other organelles of eukaryotic cells function as sites of synthesis of new proteins (ribosomes) and modification and transport of proteins to the membrane and other sites in the cell (the endoplasmic reticulum and the Golgi apparatus).

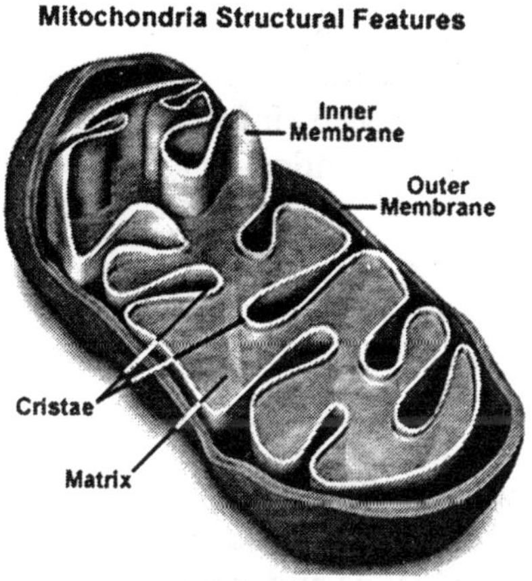

Fig. Ribosomes

Plant cells have organelles called chloro- plasts that carry out photosynthesis, the specialized biochemistry of plants that allows them to use light energy for the synthesis of sugars from carbon dioxide and water. Certain properties of chloroplasts, the double membrane and the possession of genetic material, suggest that they, as mitochondria, developed from a formerly free-living prokaryote, in this case one capable of photosynthesis.

The cell membrane of eukaryotes is composed of predominantly lipid and proteins with some sugars.

Prokaryotes have a cell wall composed of amino acids and sugars. The cytoplasm of eukaryotic cells is not simply a membrane-bound bag of organelles, but contains a complex set of structural proteins, called the cytoskeleton, which gives the cell shape and may give it specialized properties such as the ability to move, to contract or to respond to outside stimuli by changing shape.

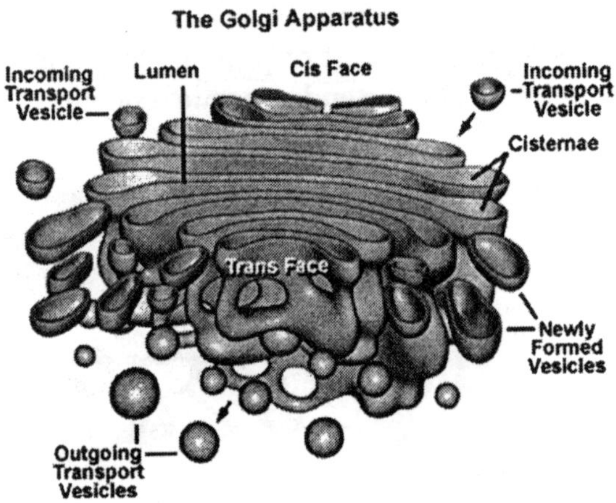

Fig. Golgi apparatus

Chromosomes and Inheritance

During the middle of the nineteenth century it was concluded that living beings could only arise from other living beings and that life did not arise from nonliving chemical substances. And it was also concluded that parents passed on to their offspring certain properties or traits. Mendel deduced the rules of this process with his studies of peas. In later decades, as the science of microscopy was advancing, the structures responsible for inheritance, chromosomes, were identified by careful observation of cells before and after cell division.

The chromosomes in the nucleus of cells reproduced (doubled) before a cell divided and were then distributed evenly into the daughter cells. This suggested that these structures, the paired chromosomes, carried the information that caused the daughter cells arising from cell division (mitosis) to resemble the parental cells. It was also observed that in fertilization, in sexual reproduction, the sperm contributed a single set of membrane-bound chromosomes to the egg, which contained a nucleus with a single set of chromosomes and that the two parental nuclei fused following fertilization.

Thus the chromosomes carry the genetic information that determined the heritable properties of the products of asexual and sexual reproduction. In the 1940s and 1950s scientists such as Tatum, Lederberg and Zinder established that bacteria exchange genetic material by pairing and then inserting the chromosome from one to another.

A phenomenon called bacterial transformation provided evidence that inanimate chemicals carried genetic information. In the 1920s and 1930s Griffith had found that dead bacteria of S strain pneumocci (toxic to a mouse if alive, but not if the bacteria were killed) and live R strain, harmless on its own, were deadly when injected together. This suggested that a chemical from the dead S strain had changed and transformed the R strain into an S strain. Avery, MacLeod and McCarty proved that the transforming material was DNA.

But the conclusion that genetic material in general was DNA was resisted by many scientists who held on to the idea that the genetic material was protein. Benzer and others showed that transformation occurred naturally through the means of bacteriophage, bacterial viruses that move genetic information from one bacterium to another. Hershey and Chase performed a clever and what now seems a simple experiment to find out if the protein or DNA of the bacteriophage were responsible for transformation.

They grew the bacteria from which they isolated the bacteriophage in medium containing an amino acid, the building block of proteins, and a nucleotide base, the building block of DNA, into which different radioactive elements had been incorporated as precursors to protein and DNA. They labeled the bacteriophage protein and DNA with different radioactive markers that could be easily detected and distinguished.

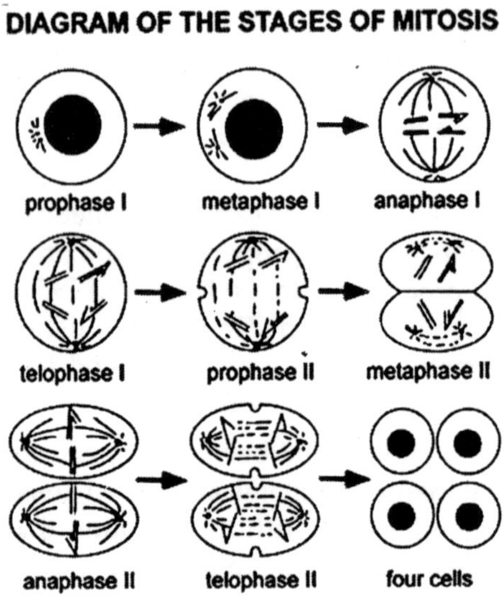

Fig. The stages of mitosis

Then, using a Waring blender to interrupt transformation at different times, they showed that genetic transformation could occur when only (labeled) DNA but not protein had entered the bacteria. This conclusively showed that for the transfer of genetic information only DNA and not protein was required. DNA carried the genetic information in transformation.

While a debate raged during the 1940s and 1950s as to the nature of the genetic material in other systems, this key work on bacterial transformation by Avery, MacLeod and McCarty and by Hershey and Chase in the 1950s supported the growing belief that DNA was the genetic material responsible not only for bacterial transformation, but in general.

Scientists began to think that DNA provided the genetic information in the details of its chemical structure. Chargaff and others had determined that DNA is a chain of small molecules called bases. There are four chemically different bases: adenine (A), guanine (G), cytosine (C) and thymidine (T). Chargaff carefully analyzed the composition of DNA and found that the number of As was equal to the number of Ts and the number of Cs equaled the number of Gs. The key seemed to be in the order, the sequence, of the bases.

The question that engaged many scientists during the late 1940s and early 1950s was how DNA could provide for a copy of itself, a requirement for the DNA to transmit the genetic information from one cell to the daughter cells at cell division.

In 1953 Watson and Crick published their historic paper showing through X-ray studies that DNA was double stranded and that the bases of each strand of DNA paired with the other. They commented in this revolutionary paper that "It has not escaped our notice that the specific pairing we have postulated immediately suggests a copying mechanism for the genetic material". For the synthesis of two copies of DNA to pass on to daughter cells, the original double strands separate and each lines up the bases of the new strand. The bases are linked by an enzyme called DNA polymerase.

The double-stranded DNA thus directs the formation of a double-stranded copy by base pairing. Though it was clear that the genetic material is DNA and it can be duplicated by base pairing, this still doesn't connect the gene to the trait. To return to Mendel's experiments, how does the gene determine seed colour? The technical and scientific keys to answer this question came from biochemistry and genetics.

Biochemists were studying enzymes, the specialized proteins that carry out the chemical reactions in cells by acting as catalysts. Much of the function and structure of cells and thus of whole organisms were seen to be determined by proteins. A key series of studies relating proteins to genes were carried out with bread mold.

Beadle and Tatum exposed bread mold to X-rays to cause mutations, isolating a series of mutants, each unable to grow unless a certain nutrient was present in the growth media. They then showed that each distinct loss in the ability to use a nutrient was related to an alteration in an enzyme involved in the metabolism of that nutrient. The change was inherited by the offspring of the irradiated mold, and the heritable differences were manifested or expressed in specific proteins in the bread mold.

Proteins are, or through their enzymatic activity determine, the functional and structural molecules of cells. As geneticists and physicists were working toward defining the nature of the genetic material, biochemists were developing tools and techniques to study the structure and function of protein. They learned that proteins were composed of chains of amino acids and that different proteins had different sequences of amino acids resulting in different functions and structures.

There are only twenty different amino acids that are strung together to make the 100,000 or so proteins found in an animal or human cell. The conclusions were clear. The combinations of amino acids are not random, but determined by the structure of the genetic material. The instructions for each protein had to be encoded in the sequence of the gene. But it still was not clear how DNA, now seen as the genetic material, codes for proteins.

With the solution of the structure of DNA by Watson and Crick and the implications of base pairing, it was reasonable to propose that the sequence of bases of DNA provided the genetic information to direct the synthesis of proteins. How could the four bases in DNA determine the choice and order among 20 amino acids in proteins? It called for a code and a decoder mechanism that would direct the cell to synthesize the protein.

From Gene to Protein

The solution to the problem of how the information in DNA resulted in the synthesis of proteins started with studies of the biochemistry of cells that were very active in making new protein, such as cells of an embryo, adult liver cells and bacteria infected by DNA viruses, bacteriophage. Analysis of such cells showed that they had unusually high amounts of a nucleic acid similar to DNA but with a different sugar (ribose instead of deoxyribose) as part of each base, and with the substitution of uracil, a base with a different structure, rather than thymidine.

Most RNA was also single, not double stranded. Cells actively making protein made RNA in the nucleus, and then the RNA moved to the cytoplasm. The DNA in the nucleus directed the synthesis of RNA, by base pairing, so that the sequence of stretches of RNA would be complementary to stretches of DNA with the uracil substituting for thymidine.

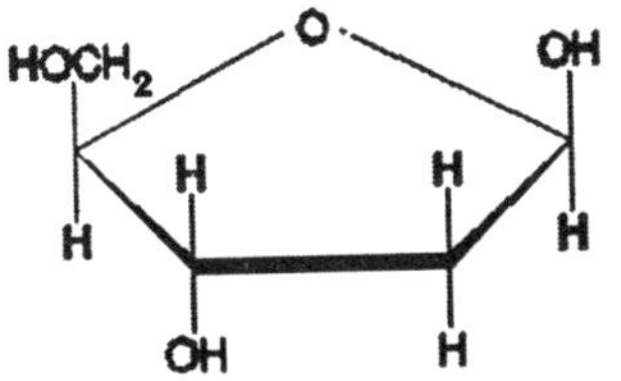

Deoxyribose

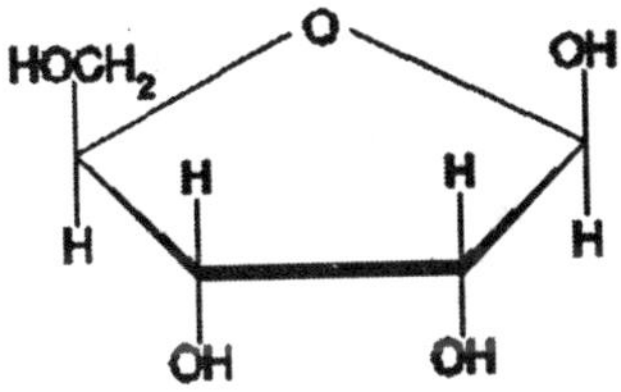

Ribose

Fig. Deoxyribose

Crick proposed a three base code to go from DNA to protein going through the RNA intermediary. The biochemistry of protein synthesis was sufficiently understood by this time so that a cell free system could be assembled with all the right enzymes and components, but no natural RNA. This cell free system would only make the polypeptide (the string of linked amino acids) coded for by the RNA provided to it. To this system Nirenberg and Khorana systematically added synthetic RNAs, first made with a single repeating base and then with sets of three bases repeated in triplets.

By analyzing the composition of the polypeptide made in this cell free system with each RNA composed of a repeating triplet, they were able to deduce the coding rules for the genetic material. The system told them the code. If RNA with a string of three cytosine (CCC) repeated gave a string of prolines, then the code in the DNA for proline was GGG, the complement for CCC. If you put a string of triplets together, you have the code for a particular amino acid sequence. With the results from these experiments Nirenberg and Khorana could compile the code for translating RNA to protein.

The process of protein synthesis is actually a bit more complicated since the machinery of protein synthesis in a cell involves three types of RNA and a number of proteins. RNA with the sequence of triplets encoding the amino acid sequence is the messenger RNA (mRNA). Proteins are synthesized on cellular structures called ribosomes, composed of specialized RNA, called ribosomal RNA (rRNA), complexed with the enzyme proteins that create the peptide bonds linking together the amino acids called for by the messenger RNA that is read on the ribosome.

The third form of RNA is call transfer RNA (tRNA), which binds to an amino acid and shuttles it (transfers it) to the ribosome. There is at least one kind of tRNA for each amino acid. The process of synthesizing a mRNA molecule from the sequence of bases in the DNA, the gene, is called transcription and is catalyzed by a protein called RNA polymerase.

Transcription is highly regulated in the cell, influenced by environmental factors and the products of other genes. Research over the last thirty years has shown that a gene is not composed simply of the bases coding for the particular protein, but also includes control elements interacting with various regulatory molecules, which are often but not always proteins. One such control element of the gene is the promoter, a docking site for the RNA polymerase, signaling where to start transcription.

The regulation of transcription is highly complex and is specific to the nature of the protein to be synthesized and its role in the cell. For example, the regulation of the synthesis of an enzyme involved in the metabolism of a particular nutrient may be responsive to the presence of that nutrient in the cell's environment.

Regulation in multicellular eukaryotic organisms' cells, with many cells different from one another, is complex, specific and not completely understood. This is the problem, as one student said, "of why my liver doesn't grow hair." It remains the focus of a great deal of research.

The process from mRNA to protein is called translation. Translation occurs on the ribosomes. Given the number of molecules involved in translation (tRNA, mRNA, rRNA and several structural and catalytic proteins), it is not surprising that this offers further opportunities for regulation, for controlling whether a particular protein is made or not. In these as yet poorly understood ways the cell orchestrates what it is made of and what functions it has.

Genes in eukaryotic cells are organized into stretches of bases that are transcribed and translated into the protein product and stretches of bases that do not end up decoded into the protein product. The segments of bases that are expressed, called exons, are interrupted by stretches of DNA that remain unexpressed, introns.

A gene may be composed of a few exons, with a few intervening introns or many exons twenty or more) and many intervening introns Some of the control elements are encoded in the introns, others are encoded at the beginning and end of the structural gene sequence.

This complex organization of the gene in which the regulators, repressors and promoters are part of the gene structure, its sequence, but do not end up translated into protein, provide many ways for the cell to control whether the gene is expressed, under what conditions and for how long. The organization of the exons and introns also provides clues to the origins of the gene, since the exons and introns may be moved about or shuffled during evolution.

The exons and introns of the gene are transcribed into messenger RNA, which is then processed to remove the RNA complementary to the introns, resulting in mature mRNA encoding for the protein product. Many proteins are also further processed after their translation by digesting off bits, adding sugars or fats to certain amino acids and modified in other ways.

These modifications can affect the function of the protein and where it ends up in the cell or in the whole animal. These posttranslational modifications may also affect how long the protein survives in the body of an organism. The biochemical ways that prokaryotes can and do modify proteins are different from the posttranslational modification in eukaryotic cells. This is significant in the use of prokaryotes such as bacteria, for expression of mammalian proteins for pharmaceutical uses.

Engineering Genes

Work during the 1960s on the biochemistry of bacterial genetics, the means by which DNA was copied, how bacteriophage integrated into the bacterial DNA and the discovery of enzymes that could allow for the synthesis of DNA molecules at will in the lab, suggested that the means for engineering bacteria, for modifying bacteria, would be at hand.

Putting Genes into Bacteria

The first effort to take a DNA sequence from one organism and put it into another in a way that allows the gene to be expressed was a demonstration done by Boyer and Cohen. They had been studying the new type of enzyme, the restriction enzyme, from E. coli, which cleaved DNA into fragments with sticky ends, because the cut is not straight across through the two strands but occurs on one strand a few bases from the other.

This leaves a tail on each piece of DNA able to pair with DNA cut by the same restriction enzyme. They applied the restriction enzymes to engineer bacterial plasmids, small circular DNA molecules found inside bacteria.

When a bacterium reproduces, the plasmids it carries also reproduce. Two different plasmids were each cut by the restriction enzyme EcoR1 and then mixed and treated with DNA ligase, which annealed the cuts of double-stranded DNA. The resulting hybrid construct could be used to transform some bacteria. Elements of both plasmids joined in this way will replicate with the bacteria.

A plasmid had been isolated with a gene for resistance to the antibiotic tetracycline, and they knew the gene for antibiotic resistance was unaffected by EcoR1. The EcoR1 enzyme could be used as a scalpel to cut out the antibiotic resistance gene and put it together with another gene into a single plasmid. The presence of the hybrid or recombinant plasmid could be identified by growing the transformed bacteria in the presence of tetracycline and testing the bacteria that grew for the other gene.

This had been done previously with plasmids isolated from different bacteria and carrying different genes, for example two different genes that would make the bacteria resistant to different antibiotics. When a hybrid plasmid with both antibiotic resistant genes was made and put into bacteria the bacteria was resistant to both antibiotics. The challenge was to use a piece of DNA not from another plasmid, but from another organism.

Methods for physically isolating the DNA for two major ribosomal RNAs of a toad had been developed by digestion with EcoR1. These fragments were mixed with a plasmid with the antibiotic resistance gene also cut by EcoR1 and the resulting mixtures, after ligation, reintroduced into the E.coli.

Several transformants were isolated after growth in tetracycline and were found to carry DNA that would bind to toad DNA fragments of the correct size for ribosomal RNA. Most importantly the DNA was expressed, that is, RNA isolated from the transformants also bound with the appropriate piece of DNA from the toad.

This experiment was not only the beginning of modern genetic engineering, but it set off a highly publicized debate about the safety and ethics of this ability to move genes around at will. This debate had its theater at a conference on recombinant DNA held in Pacific Grove, California, in 1975.

Chapter 2

Generating Research

THE EARLY TO MID-1970S

With a more complete understanding of the functioning and regulation of gene activity in development and differentiation, these processes may be more efficiently controlled and regulated, not only to avoid structural and metabolic errors in the developing organism but also to produce better organisms.

Introduction

This chapter presents historical material from the early to mid- 1970s detailing the early history of genetic engineering techniques as the basis of production for human proteins. This history starts with the development of these techniques through scientific research in universities and goes on in subsequent chapters to examine other types of knowledgeseeking activities necessary for early commercial uses. The historical period covered in this — the 1970s — was a decade with major scientific breakthroughs and with the initial recognition by individuals, firms, and other organizations that genetic engineering might be profitable for making certain pharmaceuticals on a large commercial scale.

The scientific results enabling genetic engineering techniques built upon more than twenty years of basic research in molecular biology, microbiology, and related fields on DNA, genes, and on cells. The genetic engineering techniques developed in the 1970s enabled controlled changes to DNA and followed, to a large extent, the logic and possibilities that molecular biologists had understood even when they had theoretical knowledge but no practical techniques. That is the meaning of the chapter epigraph, which is from 1958 and foreshadows these developments.

During the first years of the 1970s, experiments began to show that relatively simple techniques could be devised for genetic engineering — relatively simple, that is, for the skilled scientist. Control and regulation of the function of genes and thereby of organisms became practically possible. Direct medical applications of the new technology also appeared to be within the reach of science, although still in the future. Much of these developments occurred in the basic scientific environment.

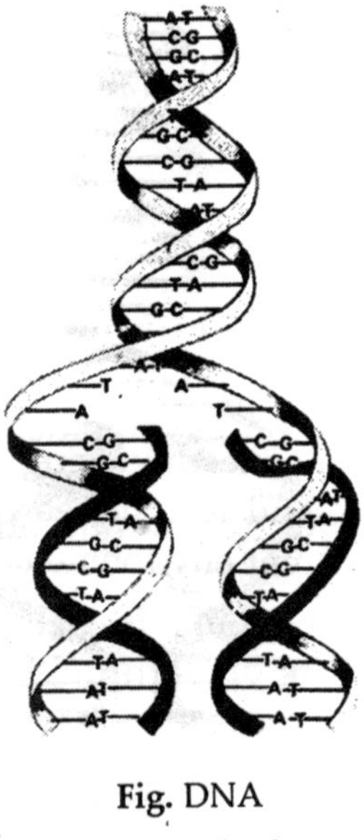

Fig. DNA

Alongside basic research unlocking the secrets of life and of molecules like DNA, another type of use of genetic engineering was beginning to develop in the 1970s, namely commercial uses of genetic engineering to produce pharmaceuticals. The 1976 founding of the Californian biotech firm Genentech is often seen as the start of this new biotech industry, where basic scientists and venture capitalists joined together. These small biotech firms generally sold R & D contracts to established companies. These contracts were to develop new scientific knowledge and techniques in order to be able to adapt what had initially been scientific activities to commercial purposes.

The American pharmaceutical firm Eli Lilly (Lilly) was quickly interested in this technology for insulin, as was the Swedish pharmaceutical firm KabiVitrum (Kabi) for human growth hormone. Their interest led to two of the first R & D contracts recognizing the economic and technical potential of genetic engineering techniques. Genentech carried out the respective R & D contracts on genetic engineering for Lilly and Kabi.

Thus, although scientific research was vital for the generation and development of genetic engineering techniques in the 1970s, it was also an important period of developing and shaping genetic engineering techniques for commercial purposes. This changed the directions and search spaces of agents' knowledge-seeking activities, and thereby changed the probable results obtained. In the early 1970s, the scientific-economic environment was just beginning to be constructed for this type of activities involving genetic engineering. The same knowledge and techniques could also be applied to advantage in different environmental conditions.

The dual uses of genetic engineering knowledge, i.e. to further the scientific endeavour and to further the profit motive, seem to fit the two categories of 'basic science' versus 'applied science'. One might even assume that during the 1970s, governments were funding scientific activities and firms were funding technological activities. They were, and this is a basic trend. In addition, however, firms were funding some scientific activities in molecular biology, and some basic science funded by governments was used to develop new technology relevant for commercial profitability.

Thus, the history of events exemplified by insulin and human growth hormone does not fit with the implicit assumption that scientific research necessarily has to reach a point where useful techniques or ideas can be handed off to industry before firms will become interested in science. Instead, as Kline and Rosenberg have argued, firms will directly fund basic science when no other sources of information are available.

Lilly, Kabi, and Genentech did so for insulin, hGH, and similar products in the 1970s because they could see the technical and economic potential of genetic engineering techniques. The firms funded scientific activities in an economically relevant direction and wanted knowledge, techniques, and biological material necessary for this new context of use.

The Recombinant DNA Controversy and Scientific Activities

In the early 1970s and based on much previous research, basic scientists developed experimental techniques which allowed them to combine bits of DNA; these were the basic genetic engineering techniques and gave scientists some control over nature. Moving specific genes had previously been a theoretical possibility, but scientists had lacked practical, functioning techniques. Genetic engineering techniques were thus generated within the basic scientific environment, which had accumulated knowledge and techniques over several decades. These scientific fields were strongly oriented towards basic science and usually had few contacts with other environments.

When genetic engineering techniques appeared practically possible, basic scientists had many questions about how well the techniques would work in practice, about the consequences of experiments, and about whether formal regulation should be devised to determine which experiments were safe in the sense of being acceptable. Their version of regulation was to be internal to the basic scientific environment, regulated by scientists based on scientific research and criteria.

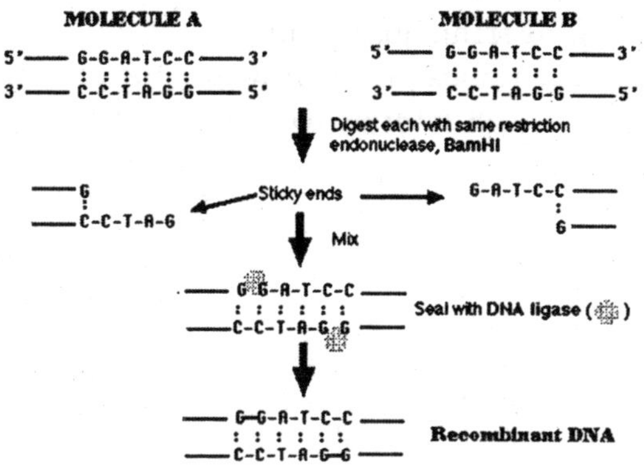

Fig. Recombinant DNA

Unusually perhaps for basic scientists, but with the nuclear physicists and their earlier appeals to stop atomic weapons development in mind, these scientists were active early on in trying to encourage debate and formal regulation thereby shaping and creating a predictable public context for the technology.

On 26 July 1974, a handful of American scientists published a letter which would subsequently influence formal regulation of genetic engineering, also called recombinant DNA (rDNA), in many countries. This started what has been dubbed the recombinant DNA controversy. Following appeals voiced at the 1973 Gordon Research Conference on nucleic acids, the American National Research Council formed a committee of eleven leading researchers to investigate the potential dangers of genetic engineering experiments.

The committee was headed by the renowned biochemist Paul Berg, who had performed early experiments indicating the feasibility of these techniques. This committee developed some recommendations to regulate experiments, including the forbidding of some, and subsequently published a letter in the influential magazine *Science*.

This was the letter that would influence much of the subsequent course of events. In it, the Berg committee proposed a voluntary, international moratorium on certain genetic engineering experiments which 'could prove biologically hazardous'. A main reason for concern was that experiments in laboratories had indicated that researchers could link sequences of DNA from different species, leading to truly new and unknown combinations.

This moratorium was internal to science, partly because only a handful of élite basic researchers had then actually mastered the genetic engineering techniques necessary to succeed with experiments. Moreover, the appeal and subsequent debate were mostly published in scientific journals like *Science*, and the regulatory guidelines which most debaters had in mind would be designed, watched over, and evaluated by scientists.

Even those scientists who felt the proposed moratorium was overly cautious felt compelled to act — especially to delay experiments — once the letter became public. The moratorium was voluntary, but in the informal institutions of the basic scientific environment, norms are often internalized. Individuals felt compelled to follow the rules and norms which were to be agreed upon collectively, even when control was informal. In the words of Professor Lennart Philipson from Sweden, who planned an early recombinant DNA experiment with Pierre Tiollais of the Pasteur Institute, France:

Once the information was out in the open, we had to wait for the reaction, and for some sort of guidelines. Our feeling was that we had to hold back until there was an agreement on how to perform the experiment.

I felt very early on, in fact, that maybe this letter, and the whole uproar that followed, was an effect of biochemists being concerned with problems not previously encountered. Medical microbiologists have been exposed to similar dangers for a long time and sort of learned to live with them. Guidelines and rules were available.

Despite early arguments among scientists about whether the Berg committee had overstated the risks, the Berg letter and subsequent debate had the desired effect of stopping certain types of experiments until the envisioned risks could be evaluated. During a period, basic scientists defined a new public context and closed off certain technological trajectories of development.

On the voluntary, informal level, the moratorium on some scientific experiments lasted until May 1975, when scientists met at the Asilomar Conference. They agreed on classification of experiments into risk groups, with corresponding requirements about physical containment. *Science* again acted as the forum to diffuse information, as a summary of the executive report was published on 6 June.

The report included guidelines for classification of experiments, and it would ultimately steer scientists' experiments for a year, until the American National Institutes of Health (NIH) established their safety guidelines in June 1976. The NIH is the major medical research council in the USA as well as the home of major medical research institutes. Their guidelines were also followed internationally in this research community, until national guidelines were developed in other countries.

Following the Asilomar recommendations, the NIH classified experiments into four risk types, called P1 to P4. The numbers run from the least to the most restrictive category, and each has specific physical and biological containment requirements for experiments. The NIH guidelines were to be followed in all publicly funded research in the United States, but did not legally apply to American industry, only morally. These guidelines were made by scientists for scientists, with little or no public insight, probably because scientists felt they could best evaluate the risks. Basic scientists did not directly attempt to regulate experiments for other environments, partly because genetic engineering techniques were then considered only of interest within the basic scientific environment.

Although the 1973 Gordon and 1975 Asilomar meetings took place in the United States, scientific research is an international activity. Those scientists asking similar research questions in Europe and elsewhere were aware of the American discussions through a range of interactions, from visits by American researchers, Europeans spending a post-doc year in the USA, international conferences, and direct participation in discussions.

For example, Philipson and others from the European Molecular Biological Labs participated in the Asilomar conference, and like American scientists, followed the Berg letter guidelines. They restricted experiments until the norms of safety in the international, basic scientific environment could be agreed upon. Moreover, scientists in many countries tried to communicate the importance of genetic engineering and its risks and benefits to society and to develop national regulation.

Despite the international character of science, scientists located in the USA made many of the relevant breakthroughs, even if their work was based on an international community of researchers. Philipson argued that 'it was clear that most of the technology, the restriction enzymes and the recombinant technique, emanated from the United States.' At the same time, much of the foundation of knowledge necessary for genetic engineering had previously been developed in Europe, and many younger European scientists worked in the United States and contributed to those scientific developments.

This cumulative nature of science must not be forgotten when evaluating the relative role of different national research systems and when discussing how knowledge is accumulated. Here, however, it is not possible to give credit to the many researchers who contributed through experiments and theoretical work. Instead, this section will take up a few of the major scientific developments relevant to the history of the commercialization of these techniques for hGH and insulin.

In addition to Berg's own research, one experiment which started the recombinant DNA controversy involved at least one basic scientist who would soon thereafter act to realise the commercial potential of genetic engineering. In 1973, two active research groups in relevant scientific fields were centred around Professors Stanley Cohen of Stanford University and Herbert Boyer of the University of California, San Francisco (UCSF). Cohen and Boyer put together a series of techniques to manipulate DNA, the so-called recombinant technique.

The recombinant technique is really a series of techniques, including cutting segments of DNA with restriction enzymes, pasting together two fragments with ligases, and getting the genes into a bacteria cell through a circular fragment called a plasmid. Not only could researchers cut and paste genes, they could do so at specific points, thereby having more precise control. Their technique was initially quite rudimentary, but it worked. In that Cohen and Boyer developed this technique in 1973, useful and relatively simple techniques were only just being developed when the international, internal scientific moratorium was called in 1974.

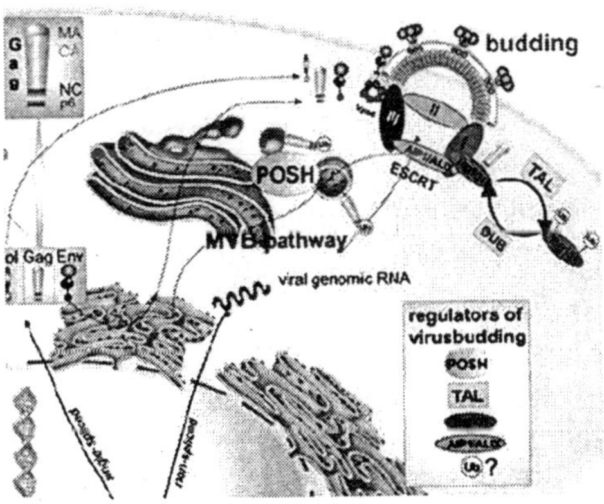

Fig. Ligases

With the establishment of, first, the informal guidelines in 1975, and then the formal ones by the NIH in 1976 (copied by a number of countries), these basic scientists could get back to research which used or developed genetic engineering techniques. In the mid-1970s, they had a handful of techniques and a handful of powerful theories about DNA, its universal character, and its importance in instructing cells to make proteins. The simplest type of genetic engineering is inserting a bit of DNA into a single-cell bacterium like *E. coli.* This had only recently become possible and was only relatively 'simple' in the sense of being able to do without huge mobilizations of labour and resources. Scientists still had to be trained to utilize these techniques.

What these basic scientists did not have in the mid- 1970s was the certainty that the mixing of techniques, biological materials, and theories would lead in practice to the results to which scientists thought they ought to lead. Theories had not been confirmed. It might not work. Despite relative agreement over theories, scientists experimented with different approaches to practical techniques, partly because no one was sure which path would prove most fruitful.

We can say that uncertainty about which experimental approaches would work in practice encouraged competing scientific research groups to generate diversity. Moreover, because conditions in the basic scientific environment include incentive structures which reward those scientists who publish first and accurately, competing groups had incentives to spend resources to try out quickly alternatives which seemed most promising. Speed and high quality in relation to internal scientific criteria were encouraged.

Scientists developing these techniques also faced a series of unanswered questions about the viability of genetic engineering for actually expressing human proteins. In particular, in the early to mid-1970s, no one had yet proved the next step after cutting and pasting DNA segments, namely whether or not bacteria cells could actually 'read' a gene as an instruction to make a specific human protein. Even before the moratorium, various scientists were considering experiments to explore this possibility. Until shown in practice, scientists remained uncertain about whether or not this would be possible.

Theories predicted it should be possible, but questions remained in a number of dimensions. Could a'bacteria build proteins normally only found in humans? Would the DNA sequence stay inside the bacteria or would the bacteria reject it and revert to what is known as a 'wild type'? How could scientists build their own DNA sequences in the lab? By using the fundamental chemical components, they would no longer have to rely on DNA already existing in cells. Even if the bacteria succeeded in producing the desired protein, would the protein keep its chemical and physical structure over time? Would it be folded correctly? Would other parts of the bacteria (immediately) attack and destroy the protein in the cell? Expression should work if DNA was as universal as theory predicted, but it was difficult to prove and required the development of experiments and techniques. These unanswered questions will also be used later to help explain why experiments with human growth hormone in 1979 were considered scientifically — as well as commercially - interesting.

As mentioned above, pioneering work on the rDNA technique was developed in the early 1970s at the élite, private Stanford University and at the California state-funded UCSF, which had research of special renown in medical fields. These two universities, as well as the California state-funded University of California, Berkeley, all did significant research leading to genetic engineering techniques. They all lie in or nearby San Francisco, California, in the so-called Bay Area, within an eighty-kilometre (fifty-mile) radius.

Even though science is international, this local university climate should be kept in mind as the focus of this narrows more and more to specific individual researchers, university departments, biotech firms, and pharmaceuticals firms which contributed to the use of genetic engineering techniques for insulin and human growth hormone. For, if we jump from 1976 to 1979, we see that not one but two research groups in the Bay Area published scientific papers in respected journals about how to use bacteria to produce hGH.

Moreover, one group was situated at a university, UCSF, and the other group at the biotech firm, Genentech. Both organizations had active basic researchers; both generated basic scientific results; and both carried out research financed by established pharmaceutical firms. Concentration of researchers in one geographical area, in this case, did not entail pooling of resources to perform one joint project but instead entailed competing groups. In fact, there were conflicts as well as co-operation between these groups.

The explicit ties between UCSF and Genentech are closer than first meets the eye. To uncover the tangled relationships, the remainder of this moves chronologically from 1976 to 1979, starting in a university department and drawing in other individuals, firms, and organizations as their activities touch upon the technological innovation process for hGH. In this way, the factors and mechanisms underlying the pattern of coevolutionary innovation are highlighted from the perspective of both basic scientists and corporate researchers. The implications should therefore be of interest to government policymakers, universities, and firms alike.

Inside UCSF, scientists in several departments worked on research questions that either directly or tangentially enabled further development of genetic engineering techniques. Dr William Rutter, head of the Department of Biochemistry and Biophysics since 1969, was building up the department's strength in basic research by attracting researchers who had ideas and the ability to pull in research grants and who could develop labs with fancy and mundane equipment.

One of the three senior researchers in Rutter's department who would eventually be co-author on a scientific paper applying genetic engineering to human growth hormone was Howard M. Goodman. A second was John D. Baxter, a medical doctor with his major appointment in medicine, but who sometimes worked closely with Goodman. These were the last two (out of four) authors on the UCSF paper on hGH published in 1979. A third UCSF faculty member, Herbert Boyer, also had an appointment in this department but had his major appointment in microbiology. Boyer is particularly interesting because he started the biotech firm Genentech, which also published an article on human growth hormone in 1979.

Thus, one department at one university sufficed to generate two different attempts to answer almost the same research question. It is therefore interesting to know more about how the informal institutions of basic science worked to stimulate this diversity. Moreover, we can ask whether these individuals were responding to the same environments or whether both the basic scientific and the scientific-economic environments were important.

An understanding of how Goodman, Baxter, and Boyer came to be doing hGH research in the late 1970s should be based on their previous research projects. In 1975, Baxter was working on medical questions about hypertension whereas Goodman was clearly working in a field related to genetic engineering, in his third year of NIH support for the project 'Structure and function of tumour virus DNA SV40'. Moreover, Goodman had previously worked intimately with Boyer on a number of experiments. Boyer had developed restriction enzymes and sequenced the nucleotides left after the enzyme Eco RI cut DNA at specific points. The scientific research of the last two thus involved both knowledge and techniques enabling genetic engineering to work.

Because these three senior researchers co-operated in different constellations over time, it is interesting to ask why Goodman and Baxter competed with Boyer over hGH in 1979. This raises the larger question of how a scientific field which is just beginning to be perceived as commercially interesting experiences tensions beneficial, or destructive, to the creation of diversity. In this case, the tensions relate to activities which are potentially relevant for more than one environment.

One reason for tensions has to do with perceptions about the proper relationships between knowledge developed in a university and in a firm. To different extents in different fields and different times, university researchers have had contacts with firms. Individual scientists have been active in both scientific and commercial development of a number of technologies. However, when a research community has not traditionally had relationships with firms and has instead been a more isolated epistemological community, then new problems and discussions arise when their scientific activities begin to be perceived as being of economic interest.

Such tensions are relevant here because within this very same department, another trend affecting knowledge generation was simultaneously getting under way, namely commercialization of genetic engineering. In late 1975, the venture capitalist Robert Swanson discussed commercialization possibilities with the basic scientist Boyer, and this led to their forming a partnership in January 1976 and incorporating a company in April. That company is the first dedicated biotech firm, named ' Genentech'. At the time, Boyer was a tenured faculty member at UCSF and remained so for many years, but he simultaneously consulted for Genentech and helped steer its early research.

In the literally hundreds of articles written about Genentech, it is often not the scientific side — and hence the role of Boyer and other scientists-which is emphasized. Instead Swanson, the venture capitalist founder, gets the credit for perceiving the commercial potential of genetic engineering, signalling the start of a whole new industry. Swanson was no outsider to the business. He had both scientific training in chemistry, with a Bachelor of Science from Massachusetts Institute of Technology (MIT), and knowledge of the commercial world, with a Masters of Business Administration from MIT.

Swanson also had experience from working first at Citibank 'specialising in setting up small companies with venture capital' and then in 1974 at Kleiner and Perkins in California, a small venture capital company. Swanson personally had the potential to understand both the scientific-technical and economic potential of genetic engineering, as did Boyer, but their competences were largely complementary in that Swanson was specialized on the economic and financing side and Boyer on the scientific side. Swanson might recognize the potential of the science but could not carry it out.

Swanson also had a strong vision for the new biotech firm, which helped form some of their early decisions: in the long run, Genentech would not only do contract R & D but also move into manufacturing.

For Genentech, the strategy from the very beginning was to be able to make and sell our own products. Therefore, we need not only to generate the products but to manufacture them and build the marketing force to sell them.

This vision guided many of their early strategic decisions about R & D.

In order to finance a longer-term diversification from R & D contracts into manufacturing, Genentech had to have both short- and long-term goals and enough short-term financing to reach the longer-term goals. In Genentech's case, this initially translated into attracting money from commercial R & D contracts as well as capital from private investors, including Swanson's old employer Kleiner and Perkins.

This money would give them a foundation to build up competences in house and to move into commercial production of a product. In the short run, however, Genentech had to prove its basic scientific prowess before it could sell research in the scientific-economic environment. This was necessary because genetic engineering was largely basic science at the time, and high-quality science would presumably sell better (and function better) than science of low quality.

The founding of Genentech led to mixed reactions within the UCSF department about the proper type of relationships between universities and firms in general. For example, when it became known that Boyer was planning to start a company, a then first-year assistant professor, Keith Yamamoto, recalls that he brought up the topic at a small, informal faculty meeting. He was critical of using publicly funded research to make personal profits and so put forward a proposal to stop the firm in 1976. His criticism was ignored.

No one disputed that Boyer's research had been to a large extent publicly funded, mainly through NIH, although Boyer had other sources of funding like the private foundation Howard Hughes Medical Institute and contracts from pharmaceutical companies. No one disputed that the knowledge and skills that Boyer had personally acquired were largely financed by public money, or that he was likely to use his competences to benefit the firm — and by extension, benefit himself. He could help select scientists for the new firm, use his research abilities to judge and direct projects, transmit some of his competence to the firm's researchers, and so forth.

Basically, though, there were no rules against starting a firm within the university and indeed, the university had an interest in industrial relations if contract money accrued to support further research at the university. Moreover, scientists in this field had very different opinions about the proper form of relationships between university and firms. Many of the senior scientists at UCSF (and elsewhere) later became involved in starting new biotech firms, and others began working for firms, either moving there or remaining at a university but acting as consultants. UCSF as a whole had no objections at the time about Boyer starting Genentech. Tensions nevertheless arose.

In the first two years, from 1976 to 1978, Genentech existed as a legal company, but it had no labs or location of its own. Genentech instead signed contracts with researchers in universities and research institutes to perform scientific activities. Of course, Genentech did not ask just anyone in the field. For its first project on somatostatin, the firm signed an R & D contract with a pioneering scientist, who also happened to be a co-founder, Boyer. The very fact that this contract was signed indicates that both partners found the arrangement satisfactory in 1976-7.

As said in a report from a later investigation by the UCSF Academic Senate, Committee on Rules and Jurisdiction:

Boyer founded a company, Genentech, of which he is Secretary, a member of the Board, a major stockholder, and a paid consultant. The company raised money and then let a contract to the university for work to be performed in Professor Boyer's laboratory for initial experiments to incorporate structures of DNA known to code for the desired proteins into plasmids so that the human gene products could be manufactured in bacteria.

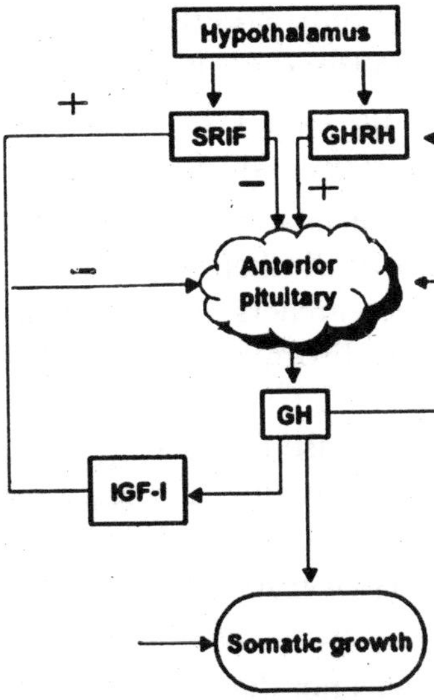

Fig. somatostatin

The principal representing Genentech in contractual matters was Robert Swanson, Genentech's President. The principal representing the university was Professor Boyer.

An exclusive licence to use any patentable material was given by the university to Genentech. Because Genentech did not have any operable laboratories of its own, initially the experiments were carried out by Professor Boyer and two postdoctoral fellows in laboratories within the department.

The relationship between Genentech and UCSF legalized in this contract thus boiled down to a relationship between Boyer and Swanson. Genentech directly drew upon UCSF researchers and facilities for its scientific activities which, in the long run, would have technological goals of manufacturing human proteins in bacteria.

This research contract gave Genentech the right to negotiate a licence with UCSF in the future, and inventions and research samples remained the property of the university. A licensing agreement between the two was signed in 1980, resulting in the payment of 'substantial royalties from the sales of human insulin and human growth hormone.' The university thus gained money not only directly for the R & D contract but also in the long run from royalty payments.

Boyer worked on this project, as did two former post-docs from the department. Although Boyer and the two former post-docs, Herbert Heyneker and Francisco Bolivar, were from the UCSF Department of Biochemistry and Biophysics, they worked closely with researchers at a large medical research hospital in southern California named 'The City of Hope National Medical Centre'. City of Hope researchers were crucial to the project because they had special competences, particularly in relation to synthetic DNA. Carrying out this initial project required cooperation between researchers with different skills, in this case located at different places in California.

Boyer, however, was not the only scientist at UCSF who was interested in putting together genes, plasmids, and bacteria. Other scientists there were interested in developing scientific knowledge and, as a by-product, expressing human proteins. Co-operating, and rival, labs and research groups were never far away. For example, by 1977, Goodman was dealing more generally with this same class of problems, and Baxter had four NIH grants, all of which dealt with medical questions relating to hormones, but two of which were tangential to genetic engineering techniques.

As another indication, Baxter and Goodman are listed as co-authors on a series of important scientific papers about genetic engineering published in 1977. Labs and experiments rely on the expertise and efforts of many individuals, and the first name on a paper generally means most work done. Therefore, it must be noted that the first authors on these papers were neither Baxter nor Goodman, but post-doc researchers, including Peter Seeburg, Axel Ullrich, and John Shine. It is praxis for senior researchers to be listed on the research of post-docs in these fields because the lab directors provide learning opportunities, ideas, advice, and the facilities for experiments.

Like the Boyer *et al.* group at UCSF and City of Hope, the constellation of researchers around Baxter and Goodman at UCSF were trying to answer some basic research questions by using genetic engineering techniques. They were testing the limits of the possible by putting a rat insulin gene and a rat growth hormone gene into bacteria, but the value in the basic scientific environment was not in making the protein but in finding the correct mammalian genes.

In May 1977, the UCSF group reported success in isolating the rat insulin gene and in inserting the gene into bacteria, but the bacteria was not then able to use that DNA fragment to make mammalian insulin. Although the scientists gained knowledge about the correct DNA sequence due to this experiment, it still left the practical question, was it possible to express a human protein in a bacteria? Their work was picked up by the press, and added to the public debate over genetic engineering. In general, Baxter's and Goodman's research groups at UCSF were quite active in developing new procedures and experiments. This constellation of scientists was clearly of the quality to compete in the international basic scientific community.

Later in 1977, the Boyer *et al.* group, on contract money from Genentech, showed that the techniques also worked in practice. Bacteria could assemble human proteins! The Boyer group published a paper in Science on 9 December 1977, about making a short, fourteen-amino-acid-long, human protein called somatostatin in *E. coli* bacteria. The list of authors included, first, four from the City of Hope hospital, and then the three from UCSF.

The City of Hope researchers played a key role in this experiment. Their contribution to this and later projects helped

Genentech succeed with developing techniques to make synthetic DNA. In particular, according to John Chirgwin, then a post-doc at UCSF, the City of Hope researcher Itakura developed the practical synthesis of small DNAs in vitro. Khorana at MIT, with whom Itakura trained, worked out the basics and synthesized a complete tRNA gene. Like the Chinese insulin, this [Khorana's method] was a tour de force but utterly impractical.

Genentech had the foresight to embrace practical synthetic DNA [as developed] by Itakura] before almost anyone in academia did. It served them well at the time, gave them tools that most universities could not provide.

Synthetic DNA has thus been important for commercial uses. Genentech's scientific activities were relevant not only for the basic scientific environment but were also relevant for the techno-economic environment, because they developed techniques to control nature. Practical techniques are often necessary in scientific experiments but are crucial for commercial uses. It is also a clear example of how basic scientists help develop knowledge and skills in their students, thus transmitting specialized knowledge and techniques among the population. The Genentech scientist Goeddel has pointed out, however, that this quote is incorrect, in that Itakura actually trained with Narang, who had in turn trained with Khorana.

By developing practical techniques to make synthetic DNA, the Genentech scientists could construct their own DNA, rather than pulling it out of a cell. Thus, if they knew in advance which sequence they required, they could make a copy chemically. They gained control over nature, the main purpose of technological activities.

As to the experiment as a whole, on the one hand, the outcome of this experiment — namely somatostatin — can be seen as a hormone without medical value or known market value. Somatostatin was much shorter than the human proteins known to have medical value, indicating that larger proteins might present more problems in expression. Moreover, the experimental test results which indicated that somatostatin was expressed in bacteria did not give much information, only a band on a gel.

The practical value of this experiment could thus be called into question. On the other hand, the experiments did show that some output of human protein was produced in bacteria, thereby showing that genetic engineering techniques were feasible. In this way, somatostatin was a test case, making it easier to perceive that genetic engineering might also work for longer proteins like insulin and growth hormone. Boyer et al.'s experiment did indicate the feasibility of this technique and also opened up important relationships for Genentech with agents in other environments than that of the basic scientific.

Although the 1977 experiment developed what was in many ways a very rudimentary technique, and only partially allayed scientists' uncertainty, it did indicate important potential trajectories for future developments within different environments. The results opened up new search spaces both for scientific activities and for technological ones.

Having a positive answer about the feasibility of the technology reduced uncertainty and thereby allowed the scientists to ask new questions and try new combinations of techniques, knowledge, and biological material. It likewise reduced uncertainty for firms about their intended commercial uses, by developing technological activities to control nature. As the 1977 scientific experiment showed the potential feasibility and practicability of genetic engineering techniques for making human proteins; several firms acted on their perceptions of the technical and economic potential of this new technology.

The technology offered new potential ways of competing. Established pharmaceutical firms moving in early wanted genetic engineering as an alternative source for making existing products, whereas new biotech firms wanted to sell scientific knowledge and possibly also use it to make pharmaceutical products with niche markets.

News about Research

Although the Genentech-financed somatostatin experiment was seen as important within the basic scientific environment, it also had consequences for firms and for the public context. So far, these had been only potential commercial and medical benefits which did not exist as such; they had been perceptions that the involved agents actively tried to create. In cases of radical technological change like genetic engineering, perceptions are vital to shaping political, economic, scientific, and technological environmental conditions. Somatostatin was clearly used for such a purpose.

Scientific results are usually presented at scientific conferences prior to publication but are usually not announced in the mass media before publication. News about somatostatin followed a different pattern, and its publicity was clearly used to shape the public context in the United States.

News of the experimental success with expressing a human protein in bacteria did not wait until its scientific publication on 9 December 1977. It was hot news! A full month earlier, the somatostatin results were made public at a Senate subcommittee hearing, and influential newspapers like the *Washington Post* and *New York Times* published their accounts. The *San Francisco Chronicle* also ran several stories in November and December 1977. In these, Boyer is sometimes identified as a UCSF researcher and sometimes as co-founder of Genentech, but not both at once. By neglecting Boyer's dual role, this indicates that public debate differentiated more strongly between scientific and technological activities than the agents themselves.

A major reason why the press identified somatostatin as important scientific research was that Philip Handler, who was then president of the National Academy of Sciences, leaked the news to a Senate subcommittee hearing. He called it a scientific triumph of the first order. Paul Berg, the leading Stanford molecular biologist and previously head of the Berg committee, similarly testified by stressing the scientific importance of this experiment. Regardless of the scientific significance of the somatostatin experiment, its publicity was also a useful trump card in the political pack.

This Senate hearing was held when the US Congress was debating national regulation of genetic engineering in both firms and universities. Scientists were wary that these political attempts to shape the public context would not allow them to continue with their knowledge-seeking activities. 'Dr. Handler told the subcommittee he believed that the gene-splicing risks did not exceed those of everyday, unregulated, routine hospital practice. He urged Congress to hold off passing a DNA-control law until more is known.' The scientists were thus very positive about Genentech's results.

During 1976 and 1977, more and more scientists – including those who had previously supported the moratorium and regulation – began to feel the risks were being overstated in the public debate. They were concerned because at the same time that new experiments indicated the safety level was higher than previously assumed, the mass media were increasingly critical of genetic engineering in general. Scientists were worried that if national legislation were passed, their informal moratorium would turn into a monster, leading to strict regulations.

In particular, Senator Edward Kennedy's proposed bill to the US Congress would have imposed harsh penalties and strictly regulated genetic engineering research, to a level unacceptable to scientists accustomed to freedom of enquiry. Therefore, many scientists worked to convince Congress not to legislate, and they were either active individually or organized in, for example, the American Society for Microbiology. Amidst this political negotiating, research which showed the positive medical benefits of genetic engineering and which could cool the legislative zest was welcomed.

Positive publicity about the medical benefits of genetic engineering did help researchers modify the public context. The US Congress did not pass legislation, even though most contemporary observers had thought they would. Instead, the

NIH rules continued to be in place, to impose guidelines for certain experiments in universities.

Thus, on one level, the somatostatin results came at a time when American scientists were becoming concerned that the public debate would impose harsher regulations than the scientists thought necessary. Basic scientists were therefore interested in changing the attitude of the public debate from negative to positive. The Swedish debate lagged behind the American debate, and somatostatin was not important in the Swedish debate.

On another level, Genentech itself was interested in getting publicity for somatostatin because this result indicated that genetic engineering could be used practically — and eventually commercially. More importantly for the firm, it indicated that Genentech could deliver such commercially relevant scientific results. On that level, somatostatin was important for Kabi's subsequent actions of contacting Genentech. Knowledge-seeking activities in response to one environment could thus influence knowledge-seeking activities in another environment as well as the public context.

This discussion of somatostatin and of legislation should be related to the ongoing debates over genetic engineering. To the wider public in the United States and in Sweden, the scientists' own fears, research questions, and previous debates over the technology were unclear, and their experiments were often represented as mysterious attempts to manipulate nature.

Many were uncertain about what genetic engineering techniques would and would not make possible. As one journalist said in retrospect about the United States, 1976 was not a particularly propitious time for the start-up of a genetic engineering company. The headlines were heralding monsters, not miracles Public concern about possible epidemics of cloned killer viruses prompted the city of Cambridge, Mass., to delay DNA work at Harvard and M.I.T. for months.

According to a survey on how American newspapers and news magazines reported on genetic engineering between 1976 and 1979, the height of the American mass media controversy over recombinant DNA occurred between late 1976 and mid-1977, after which the articles began stressing the benefits of genetic engineering more often and omitting the views of critics. By mid-to late 1977, scientists were beginning to argue and to show with experiments that it ought to be possible to make useful medical substances like insulin. Positive public opinion about genetic engineering techniques meant that calls in the public debate for stricter formal regulation faded away in the United States.

In July 1974, the Berg committee letter was published, and the fears and risks expressed there were taken seriously by Swedish scientists as part of the international community. Important work to implement regulation in Sweden was also taken by Professor Lennart Philipson and by Professor Peter Reichard, who sat on the Swedish Science Research Council.

A core of concerned scientists arranged meetings to spread information about genetic engineering in Sweden and to discuss the potential need to regulate its development. The first meeting was held in December 1974 between scientists, politicians, and some industrial representatives. Professors Lennart Philipson, Peter Reichard, Nils Ringertz, and Tage Eriksson informed Forskningsberedningen, the scientific advisory board to the Swedish government, about recombinant DNA. Along with scientists and politicians, the board included several representatives from industry, including Dr Bertil Aberg, a medical doctor and professor of clinical chemistry who had moved to the pharmaceutical firm Kabi. No regulatory initiatives were taken by the Swedish government at this time. The scientists felt regulation of the technology was up to them for the moment.

On the initiative of Professor Reichard, the Science Research Council formed a Genetic Manipulation Advisory Group Committee (hereafter GMAC) in 1975, which was also approved by the Medical Research Council. Reichard again approached the government, this time the Minister of Education Bertil Zachrisson, and suggested he appoint one or more members, but Zachrisson declined. This committee was similar to the NIH committee which reviewed research grants using genetic engineering techniques.

In contrast to the American NIH committee, which was constituted by scientists, the Swedish committee brought together scientists and a range of interested parties such as trade unionists. This corporatist pattern of participation is common in Swedish politics. The committee had its first official meeting in September 1976, with Professor Reichard the chairman. It had two main functions: diffusing information about rDNA to the public and to represented organizations, and reviewing research grants. The Swedish research councils had agreed only to fund experiments which had undergone review. The committee was thereby given some control over the direction of experimental developments, following accepted international praxis.

In addition to politicians and corporatist partners, the industrial sector most likely to be initially affected by genetic engineering also sat on this GMAC committee, namely pharmaceutical firms. Although none of the Swedish pharmaceutical companies had in-house R & D in molecular biology in the late 1970s, the Swedish scientists working for regulation asked the pharmaceutical association to participate. The Swedish association of pharmaceutical companies, *Lakemedelsindustriforeningen* (LIF), was thus involved in the first attempts in Sweden to regulate this technology. This indicates that scientists clearly saw that genetic engineering could soon be used by commercial firms.

LIF initially appointed the head of research of Astra, another large pharmaceutical firm, as their representative; later, that person was Aberg from Kabi. In 1976, the representative, 'on behalf of the pharmaceutical companies, agreed to the guidelines that we finally adopted, the procedure we had to review any research being done in this area, whether in a pharmaceutical industry or in a university'.

In other words, the pharmaceutical association agreed in 1976 to abide by the GMAC rules, which were informal, without direct government legislation. Their representation on the committee also gave Swedish firms early information about the possibilities of recombinant techniques. GMAC can thus be seen both as a non-government, regulatory body and an organization diffusing information about the potential of the technology.

Just as Swedish scientists participated in and reacted to the calls in 1974 of scientists working in the USA to impose a temporary moratorium, so too the Swedish media reacted to and published articles similar to American articles. *Dagens Nyheter*, one of Sweden's most influential and respected newspapers, published the article "Lord Knows what Monsters you have in your Test Tubes" on 6 March 1977. The article first sets up the views of a Harvard biochemist against the mayor of Cambridge, Mass., then discredit the biochemist while taking the fears of the mayor seriously:

> In the closed world of science, the debate about the new and forward-moving boundaries of genetic research has been going on for about five years.
>
> But the eyes of the public were not opened until Al Vellucci [mayor of Cambridge] sounded the alarm last year.
>
> One day when he heard about what went on in the laboratories in his city, he said straight out, 'Damn it, I would just like to know what monsters are going to crawl out of those damn researchers' test tubes.'

According to the tone of this article, genetic engineering was definitely dangerous and in need of political control. One of the final sentences of this article was that 'society must take the final responsibility for the actions of individual members.' This illustrates a recurrent theme among Swedish critics of the technology: not only should the Swedish state take responsibility for the direction of scientific and technological activities, it should also tightly control these dangerous scientists, who would otherwise become real-life Frankensteins. Scientists should not be left to regulate themselves.

As in the United States, authors of Swedish debate articles can be divided into proponents and opponents of genetic engineering, where both often wished to use their arguments to influence national regulation. In mid-June 1977, *Dagens Nyheter,* as the first Swedish paper to do so, ran a series of debate articles on genetic engineering. The first, written by the Social Democrat and active debater Nordal Akerman, was entitled, 'What Right does Research have to be Free?' Akerman's implicit answer in the article was 'very little'.

The second, written by Lars Wieslander, a medical researcher at the Karolinska Institute, reacted to the implication that there was 'any connection between Harvard's genetic research and monsters that people had seen on the streets!' and instead detailed the scientific basics of genetic engineering in his article. The final article in this series, 'This Concerns Survival Itself', again written by Akerman, restated the need for social control.

The articles by these two authors — a critical, professional public debater and a medical researcher — reflect an on-going dilemma in the debates about genetic engineering. For the medical and natural-sciences researchers, genetic engineering has been something understandable, techniques that can be defined and explained. For many of the debaters and the public, genetic engineering has meant possibilities and dangers, something which will be developed and used in unforeseen ways in the future. Communication becomes difficult because of these differences in basic assumptions.

By mid- November 1977 and running through spring 1978, the Swedish debate intensified, especially the negative side, which meant that Sweden could be perceived as one of the countries hostile to developments in genetic engineering. Akerman again set the negative tone in "Stop the High-Risk Laboratory!" in *Dagens Nyheter,* but other news- papers also ran articles. The catalyst for debate this time was the plans of the Biomedicinskt centrum Uppsala University to build a P3 highsecurity lab inside the university city of Uppsala.

Due to the intensity of these debates, but especially due to a decision by Uppsala City Department of Health (Uppsala kommun halsovardsnamnd), these building plans were shelved at that time. Ulf Pettersson, one of the researchers working with Professor Philipson, then carried out some experiments at the Pasteur Institute, Paris, France. This incident happened at about the same time that American scientists and public debate were praising the benefits of genetic engineering as shown in the somatostatin results.

In other words, just as the American media began emphasizing the benefits of genetic engineering and celebrating somatostatin, the Swedish media — particularly the debate articles — were emphasizing its risks. In the late 1970s, the United States offered a more favourable public context than Sweden for both scientific and technological activities involving genetic engineering.

Because genetic engineering was a radically new technology giving rise to much debate and to various proposals for regulation, the public debate played an important role in shaping national public contexts which would ultimately allow — or forbid -certain types of genetic engineering techniques and experiments.

Early Commercial Interest

One of the first firms in the world to approach Genentech about an R & D contract was the Swedish, government-owned pharmaceutical company Kabi. Kabi moved to investigate the potential of these new techniques for producing their main product quite soon after Genentech's somatostatin experiment. Through their involvement in GMAC and through in-house research groups, some of Kabi's researchers and management had heard about genetic engineering much earlier than the Swedish public.

Kabi's decision to access the new knowledge and technology through Genentech rather than in-house will make more sense if we first examine Kabi's historical involvement with pituitary hGH as interactions among corporate and university researchers, with the Swedish government playing a role. Kabi researchers had had contacts with Professors Luft and Hall at the medical research institute, Karolinska Institute. In fact, Kabi's relationships with the nearby universities and medical institute had been important for the company's whole involvement in hGH extracted from pituitary glands.

In the late 1960s, Professor Luft had, in association with assistant professor Paul Roos, suggested that Kabi make hGH. Kabi began working on pituitary hGH in 1967, started clinical tests in conjunction with Luft in 1968, and registered the product in 1971. This seems like a clear-cut example of university researchers coming up with an idea for a product which pharmaceutical firms then put into practice after having seen its economic value.

However, Kabi's initial involvement with pituitary hGH was not quite as straightforward as this depiction may seem, mostly because of firmlevel decisions being countered by issues related to its government ownership. In 1969-70, Kabi's upper management almost decided not to develop pituitary hGH as a pharmaceutical product because several market surveys indicated that a pharmaceutical designed for six or seven persons per million was uninteresting and unimportant as a product. The contemporary company directors seriously questioned whether Kabi should enter this market.

Management thought the market was quite simply too small to be interesting, but the government stepped in. As Aberg tells the continuation of this story: 'Through the intervention of the then Minister of Industry, Krister Wickman, manufacturing of hGH continued, however, and today [1982] the hGH preparation Crescormon is KabiVitrum concern's most profitable product.' In 1981, Kabi sold Crescormon for seventy million Swedish crowns per year, or about $10 million. The government owners ended up supporting a product which was much more profitable than the firm's management could initially imagine.

It is clear that Kabi's initial and continued involvement in proteins extracted from pituitary glands directly benefited from contacts with Swedish basic researchers and from government support. This indicates interesting hypotheses, such as that the reasons firms become involved with a technology may be quite local and not necessarily based on market assessment. Moreover, it indicates that if a person has power to finance a risky technological project, then personal, political, or long term visions can override decisions based on short-term market prognoses. This seems necessary for technological changes involving major technical or economic uncertainty, or both at once.

In addition to the GMAC committee, Kabi kept itself somewhat up to date with scientific activities involving genetic engineering. They were not specialists in molecular biology but heard about Genentech's success with the somatostatin experiment. According to Hans Sievertsson, who was then Kabi's research director, Kabi had a small research group of about five people who studied the complex of proteins released by the pituitary gland and the hypothalamus.

They studied interactions between hGH and other proteins, among other things. Sievertsson was a member of this group and had a Ph.D. for research on hormones in the hypothalamus. Like all research groups, these corporate researchers monitored scientific developments in the field. They could monitor these particular experiments and understand their significance, without being able to duplicate the work in-house.

In 1977, this Kabi group thought Genentech's research on somatostatin was 'fantastically exciting'. Sievertsson recollects that it was natural for them to wonder if other proteins could be produced the same way. Based on this perception, Sievertsson called Genentech during a business trip to Chicago in late 1977. He wanted to establish contact with Genentech and to look into the possibility of an R&D contract for hGH. Upon Sievertsson's return to Stockholm and after Sievertsson had talked over matters with his boss, Bengt Karlsson, Kabi invited Swanson and Boyer, as Genentech's representatives, to Sweden.

Some research managers at Kabi thus perceived the potential of genetic engineering and acted accordingly to see whether and how these scientific activities would be applicable to corporate goals in the long run.

Aberg, as research director for KabiVitrum, was also intimately involved in these discussions. In a debate book from 1982 entitled *Safe Enough? About the Introduction of a New Technology into Sweden, Recombinant DNA,* Aberg reminisces about the course of events:

I remember how Peter Reichard, professor in biochemistry at Karolinska Institute, asked me, in connection with the awarding of the Nobel Prize to Roger Guillemin — the discoverer of somatostatin — what Kabi intended to do about the new technology. The answer was that Kabi's current research director Hans Sievertsson and I, who was research director for the group Kabi-Vitrum-ACO, had already invited Herbert Boyer to come to Stockholm and discuss the possibilities for co-operation in some of the product areas that were important for Kabi.

The meeting occurred in the first week of December 1977, and Boyer was accompanied by Robert Swanson, managing director for a newly formed Californian company, Genentech Inc. Kabi was represented by Hans Sievertsson and, from the research side of the group, myself and my closest colleague, Kerstin Sirvell, Master of Science. It was clear after the meeting that it would be possible to clone human growth hormone (somatotropin, hGH) in a coli bacteria.

Kabi and Genentech representatives thus met in Stockholm to discuss an R & D contract for genetic engineering very quickly after the somatostatin experiment had gathered attention. About the same time, the scientist Reichard encouraged Kabi to explore commercial uses of the technology, but Kabi did not mention that they had already moved to contact Genentech.

Sievertsson confirms that Boyer and Swanson came to Stockholm on 'Nobeldagen' 1977, that is, 10 December, the day the Nobel Prizes are awarded. Their fee for coming was the price of their plane tickets: Genentech was interested in finding customers to whom they could sell their R & D contracts. Another researcher omitted in Aberg's account was also present at the meeting, namely Linda Fryklund, who was then department head of the Recip Hormone Laboratory.

Note, moreover, that there were overlaps between agents responding to different environments at this time as well as a chasm between an optimism in the knowledge-seeking environments and a pessimism in the public context. First, Kabi was exploring the commercial feasibility of the technology about the same time that the Swedish media was exploring the risks. Many of the debate articles were trying to shape the public context by mentioning the dangers. As discussed above, the Swedish public climate was sceptical about scientific — much less commercial -uses of genetic engineering at this point, but that did not stop Kabi from acting on their perceptions.

Second, the basic scientific environment lay very close to the two economically oriented environments. The first overlap has to do with the dual nature of Genentech's somatostatin experiment, relevant for both basic science and commercial use. The second overlap was between different ways of producing the same proteins, e.g. tissue extraction versus genetic engineering. As Aberg points out, Drs Guillemin, Schally, and Yalows were concurrently awarded the Nobel Prize in physiology/medicine for their work on peptide hormones. Guillemin had discovered the medical significance of somatostatin and had extracted it from tissue. Scientific recognition for research on hormones like somatostatin thus occurred parallel in time and within a few kilometres of the commercial recognition of a new technology which could make similar proteins.

The scientific versus commercial demands on quantity produced with the two different technologies ultimately differed by several orders of magnitude. Both Guillemin's group and Genentech's group had made about five milligrams of somatostatin, a tiny amount, before Nobeldagen 1977. However, whereas Genentech could make it in bacteria, with the potential of replicating the experiment in thousand-litre tanks, Guillemin obtained his five milligrams by grinding up almost a million sheep brains. Genetic engineering promised control over what was an apparently unlimited supply, and large-scale production was a necessary condition for this type of commercial use.

Kabi was interested in the possibilities that the new technology offered for producing their most profitable pharmaceutical product, human growth hormone. Kabi was a so-called pure, ethical pharmaceutical company which mainly produced pharmaceuticals based on biological processes. They had no competence in molecular biology, as Genentech did, but did have complementary assets and complementary technologies.

Kabi had technical and engineering experience with, for example, separation of blood products, with complex proteins, and with controlling microbiological processes such as penicillin and streptokinase. Due to this competence and experience, Kabi retained and could transmit knowledge, techniques, and equipment developed from other biological production processes and products to new projects using relatively similar new technologies. The question was, how similar would genetic engineering prove to be?

Kabi had an alternative technology for production of hGH, namely extraction from tissue, specifically extraction from human pituitary glands. In the late 1970s, Kabi was the world's largest commercial supplier of hGH extracted from human pituitary glands (pituitary or pit hGH). The nature of the international market and of supplies was an important factor underlying Kabi's decision to move into genetic engineering.

The disease hGH is supposed to treat, hypopituitary or severe dwarfism, was then estimated to afflict six to ten persons per million. This means that in 1979, Sweden, with a population of eight million, should have had about eighty cases and the USA, with a population of 220 million, about 2,200 cases. The estimated costs of one year of treatment per patient were $5,000-$10,000 in 1979, and treatment lasts about ten years, up until puberty. Either the state medical service or other medical insurance paid for the treatment. Human growth hormone extracted from tissue thus had an established, if fairly small, market. Management could easily project that the same product made with an alternative technology like genetic engineering should have at least an equally large market.

For this product, size of market was closely linked to supply of pituitary glands. The pituitary glands were often in short supply. In the late 1970s, Kabi collected about 150,000 to 200,000 glands from around the world. On average, Kabi needed 50 pituitary glands to provide enough hGH for one patient for one year, so Kabi could provide treatment for approximately 3,000 patients per year. They had 70% of the world market, with their main international competitor being Serono, a Swiss company, but there were also non-profit, government-financed agencies like the National Pituitary Agency (NPA) in Baltimore, Maryland.

The NPA gathered 6,000 pituitary glands a year, a contrast in size to Kabi. Whereas Kabi was mainly run as an independent firm, the NPA distributed the glands to medical researchers who purified out hGH and other rare brain hormones. The NPA then in turn distributed the doses to about 1,600 children without charge. The American federal government paid for extraction and distribution as part of medical research, as they felt the market would fail to provide incentives to make this pharmaceutical.

American patients plus around 50 in Sweden, if the estimated frequency is correct. This implies that Kabi sold hGH to about 2,660 patients in other countries in the mid-1970s, indicating the truly international nature of their market. Although Kabi provided hGH to countries which had shipped them the pituitary glands, the biggest markets lay in the East, especially Japan and Australia.

Commercial firms and government-financed organizations had different understandings of the potential of expanding the market for hGH. Government-financed organizations like NPA had no incentives for, or ways of, organizing a move into the new technology because of their dependence on individual medical researchers and because of the structure of direct funding. They were given money by the government to extract hormones this way. Salvatore Raiti, director of NPA in 1979, maintained 'that all patients with diagnosed hyposomatotropism are receiving the hormone'.

In other words, too little supply was not seen to be a problem, and therefore no new production technologies were needed. Raiti's statement was made in December 1979 at the symposium where Genentech announced their success in making hGH in bacteria. Since the NPA was not moving to develop the new technology, partly due to the organizational and institutional structures, genetic engineering threatened their *raison d'etre.*

Kabi had a different understanding of the potential for expanding the market in the 1970s, and this affected their perception of the potential of genetic engineering. Kabi felt it could sell all the hGH it could make because the current supply sufficed for only the most severe cases of dwarfism. Many potential patients of this disease, hence customers, simply could not get hGH. In addition to the known and established market for dwarfism, there were indications that hGH could be used for other medical conditions like burns, Turner's syndrome, and so forth. The market could be much larger than the current one expressed through pit hGH.

Even though Kabi perceived this potential market, greatly expanding the supply of pituitary hGH was not so easy. Kabi already had international contacts to collect pituitary glands (from morgues), but they continually faced a situation where potential demand was greater than supply. Raw material from biological material was not only in limited supply, it was also difficult to deal with. Aberg later claimed that these latent sourcing needs, rather than profit incentives or availability of new knowledge, pushed pharmaceutical companies into genetic engineering. Genetic engineering offered the possibility of unlimited supply because the protein would be made in bacteria or other simple cells which would rapidly multiply, each daughter cell also making hGH.

In relation to expanding the supply, the possibilities for profits were quite clear. Kabi as a firm faced economic uncertainty about the future. Kabi was having financial difficulties and needed a boost to survive in the industry in the longer term. At the same time that pituitary hGH was Kabi's most successful product and an important generator of income, Kabi teetered at the edge of bankruptcy several times in the late seventies.

This implied that if Kabi could use a new technology to increase the supply of a known product thought to have a larger market, this strategy would facilitate the firm's survival in the longer run. It could give Kabi what is today known as dynamic competitiveness. Kabi contacted Genentech about the possibility of an R & D contract due to the technical and economic potential of the technology.

Largely because Genentech was willing to sell R & D contracts which developed genetic engineering techniques for specific applications, Kabi contacted them rather than any Swedish researcher in the field. Although there was some Swedish research in tangential fields — above all nearby in Uppsala — Swedish researchers did not work on the specific application of genetic engineering which had attracted Kabi to exploring the techniques. In this case, a firm in one country could fund scientific activities in another country in order to be able to move into a new technology. Scientific competences relevant for commercial use did not have to be available locally or nationally.

Moreover, it appears that during the first negotiations between Kabi and Genentech, few outsiders to Kabi, even within Sweden, knew about the specific discussions that Aberg and Sievertsson had begun or about Kabi's general intentions to use genetic engineering. As shown above, two of the most active Swedish researchers in the field, also actively involved in regulation, were Professors Reichard and Philipson. The above quote from Aberg indicates that Reichard was unaware of Kabi's intentions in December, and Philipson had a similar opinion in July of 1977:

I don't think any one of them [Astra, Pharmacia, Kabi] is involved in recombinant DNA research at the present time. I don't think there are any plans to use it in the foreseeable future. But of course, they would like to have an option and a procedure to follow in case they would be interested.

Of course, Philipson was right that Kabi was not involved in recombinant DNA research, or even in contact with Genentech, in July 1977: their first more formal contact with Genentech was in December of that year. Their plans did, however, come in the foreseeable future. Neither was Kabi involved in the contemporary mass media debates over the technology, which otherwise would have indicated their interest publicly.

Reasons for Kabi to keep silent about their intentions could include the hostile tone of the Swedish public debate at the time as well as the competitive advantage of moving into and starting to develop a new technology before their competitors knew what they were up to.

Following the meeting between Genentech and Kabi in December 1977, the heads of the joint Kabi companies, that is, Bengt Andrén, Bertil Aberg, and Hans Sievertsson, made a decision to work out an agreement with Genentech to carry out research on hGH. They signed a 'Letter of Intent' in December 1977, which indicated their agreement to try to come to an agreement, and the lawyers from the respective companies set out to hammer out a legal contract during 1978. Kabi thus thought it likely that Genentech could successfully carry out the research project.

The 1970s was a crucial decade for genetic engineering techniques; the early to mid-1970s was analysed in this chapter. There were knowledge-seeking activities which led to novelty enabling commercial uses of genetic engineering techniques. Initially in the early 1970s, basic scientists developed these techniques and knowledge through scientific activities intended for the basic scientific environment.

Research in a number of fields, such as molecular biology and biochemistry, had built up the accumulated knowledge foundation from which recombinant DNA techniques could be developed. This research was mostly funded by medical councils financed by governments or by private foundations, but both distributed funds according to scientific criteria. This was research internal to the basic scientific environment.

This chapter began by discussing the debate about, and actions to implement, regulation inside the basic scientific environment. This occurred when genetic engineering techniques became practically possible and relatively easy to use in the early 1970s. This first section addressed what basic scientists knew (knowledge) and what they could do (techniques). Scientists' uncertainty also indicates the level of risks involved for firms, which had perceived the potential of adapting genetic engineering techniques for use in other environments.

Because the new genetic engineering techniques were relatively easy and practical to use and because no existing regulation applied to the technology, basic scientists called for an informal moratorium on certain experiments in the early 1970s. Scientists felt they had to create and implement regulation for certain experiments in order to prevent two undesirable situations:

1. Everything goes. In a situation without regulation of a powerful technology, dangerous experiments might be attempted, leading to undesirable accidents and consequences.
2. Nothing goes. Without some boundaries defining acceptable experiments, if an accident should occur, then public debate and formal regulation could become so hostile that all experiments would be banned. Both situations were suboptimal compared to finding acceptable rules and regulation for continuing scientific activities.

Even those scientists who were convinced that the initial guidelines were overly strict felt there were reasons to consider seriously the criticisms launched against unregulated development of the technology. They also felt compelled to follow the voluntary moratorium launched by other scientists.

In the mid-1970s, the scientists involved developed guidelines at the Asilomar conference, and these guidelines were later adopted by the major financier of medical research in the United States, namely the National Institutes of Health (NIH).

As in the USA, Swedish scientists informed the Swedish government about the technology during the early 1970s, but when the government showed no inclination to regulate it, scientists implemented regulation through the Genetic Manipulation Advisory Group Committee (GMAC). This committee evaluated research grant proposals for the research councils. Whereas American regulation remained heavily represented by scientists, a broader range of interest groups were represented in the Swedish committee. Agents internal to the basic scientific environment thus shaped the public context by developing norms and regulations according to internal scientific criteria such as risk evaluation.

Following the establishment of scientific guidelines, basic scientists developed and used scientific knowledge and genetic engineering techniques in order to try to express human proteins in E. coli bacteria. Much discussion centres on the two research groups who later both worked on experiments on human growth hormone and insulin. These groups were more or less within the same university department at the University of California, San Francisco (UCSF).

UCSF thus had several research groups of high quality doing this type of research, and they sometimes competed and sometimes co-operated. Some individuals focused on scientific activities for their basic scientific environment, whereas others were more directly interested in commercial applications. Even the latter had to engage in scientific as well as technological activities, as shown by the somatostatin experiment.

Scientific knowledge for the basic scientific environment placed priority on, among other things, understanding how DNA functions in cells. Developing this knowledge required making things work in practice, and this depended on a number of practical techniques, such as identifying and cutting out the relevant gene sequence and expressing the protein. This was the basic approach enabling control over nature, which also made it possible to make pharmaceuticals using genetic engineering.

Up to the mid- 1970s, agents mainly carried out knowledge-seeking activities in response to the basic scientific environment. Their actions were regulated by debate, norms, and informal regulations within this scientific community. Some scientists, however, began orienting their knowledge-seeking activities towards other environments.

In the mid- 1970s, some agents recognized and acted upon the economic potential of genetic engineering techniques. The new technology could be relevant for other uses and for agents other than basic scientists. The first dedicated biotech firm, Genentech, was started at this time by the basic scientist Boyer and the venture capitalist Swanson. Boyer remained a professor at UCSF. Established firms invested some money in in-house research, university research, or other R & D contracts. There was sometimes commercial as well as scientific rivalry, indicating co-evolutionary trends between activities for different environments.

By 1976-7, only very rudimentary techniques had actually been developed. Not even the question which most interested firms, namely, 'Can genetic engineering techniques force bacteria to produce human proteins?', had been answered. Firms interested in this commercial potential therefore faced conditions of uncertainty. Reducing this uncertainty required relatively radical results, potentially found in a combination of scientific and technological activities.

At the same time that scientists, universities, and firms engaged in knowledge-seeking activities for their respective purposes, the public context for the technology, initially shaped by scientists, was coming under scrutiny. Debate was used to influence public opinion and often called for the creation or modification of technology regulation. On the one hand, in the mid 1970s, the NIH guidelines for genetic engineering were relaxed in the United States and subsequently so in Sweden. The NIH relaxed the severity of its guidelines after scientific experiments indicated that the danger was much lower than the initial, worst case scenarios had postulated. Scientists wanted some regulation in order to continue doing experiments but only as much as deemed necessary.

On the other hand, public debate in both the USA and Sweden increasingly discussed the risks, consequences, and ultimately, the potential of the new technology. When the debate about genetic engineering became more public after 1976 in the USA and after 1977 in Sweden, the range of fears and optimism increased, and there were suggestions to forbid all experiments with genetic engineering within a city and/or within a nation. Scientists were therefore concerned when the US Congress debated federal sanctions and regulations that would have seriously restricted some types of experiments. This change would undoubtedly have affected other countries as well. While the US Congress was debating this legislation in 1977, research results were obtained which indicated *E. coli* bacteria could express the human protein somatostatin. Publicity helped create favourable public opinion for the promised medical benefits of genetic engineering, and thereby helped put off government regulation. In order to shape the public context, scientists used scientific results indicating benefits for the public at large.

The somatostatin experiment used to influence the American public context was carried out in a university lab at UCSF and at the public research hospital City of Hope. The two groups worked together under contract from the new biotech firm Genentech. Boyer, who was one of the senior scientists at UCSF and who had developed the recombinant DNA technique along with Professor Cohen, started Genentech.

The firm Genentech was initially a mixture of venture capital and basic science, with the aim of selling R & D contracts in what has been called here the scientific-economic environment. Genentech's first R & D project, internally funded but carried out externally, was this somatostatin project. It was also important because this project led to scientific research results indicating the feasibility of making other human proteins using the new technology.

This experiment led to economically useful research results for Genentech. It helped establish the scientific legitimacy and integrity of this new biotech firm. For example, these research results prompted the established Swedish pharmaceutical firm Kabi to contact Genentech in order to investigate the possibility of an R & D contract for hGH. The American pharmaceutical firm Lilly was similarly in contact with Genentech about insulin.

Kabi's exploration of Genentech's possibility of selling R & D contracts took place during a relatively heated Swedish debate in late 1977 and early 1978. Although this national debate had many negative tones, it did not stop Swedish firms from being interested in the economic and technical potential of genetic engineering. The Swedish public debate did play a role in mobilizing opinion about genetic engineering and its regulation, but cannot be seen as a decisive element affecting technological innovation processes. In this political context, Kabi initially remained silent about their decision to contact Genentech and to explore genetic engineering.

The events in the USA and Sweden up to 1977 concerning somatostatin, public debate, and the selling of the R & D contract have thus indicated a number of things about evolutionary innovation. At this point, the development of genetic engineering techniques significant for hGH was still largely the result of scientific activities carried out within research institutes, and acclaimed by scientists as interesting.

Nevertheless, alongside this basic scientific environment, Genentech's research on somatostatin also had technological objectives in that this work helped develop practical techniques, like those for synthesizing DNA. It also served as an advertisement for Genentech's scientific potential and their willingness to sell R & D contracts to firms. Science would be made relevant for managers.

Genentech's strategy was successful in that persons in powerful positions at some established firms perceived the economic and technical potential of the new technology and of Genentech's ability to deliver. Kabi had monitored external scientific research in its internal R & D departments and had participated in associations and committees which helped diffuse information about the new technology. Kabi had economic as well as scientific and technical reasons to search for a new way to produce their most successful product. For them, genetic engineering offered an alternative to tissue extraction.

The research agreements between Genentech and Kabi and between Genentech and Eli Lilly would affect the directions in which genetic engineering techniques would be developed in the future. Clearly, genetic engineering would no longer be an exclusively academic development, nor were firms willing to wait around for research results intended for the basic scientific environment but which were also applicable within the economic environments.

Some firms instead acted to support knowledge-seeking activities in order to get access to knowledge which was otherwise not available. A combination of scientific and technological activities were necessary to develop techniques which no one was quite sure would work. Exploring the search space and technical alternatives involved uncertainty. Managers and researchers had to take the risk of trying to develop knowledge and techniques which were uncertain, and at the time, scientific activities were the only way of doing so.

Chapter 3

Human Growth Hormone

MULTIPLE USES

The knowledge-seeking activities have highlighted basic scientific research for genetic engineering techniques as well as its integration into production systems. This chapter examines interdependencies between knowledge-seeking activities and agents setting environmental conditions, here over the definitions of multiple uses of, and markets for, the pharmaceutical human growth hormone. Pituitary hGH was approved to treat a condition called hypopituitary dwarfism, i.e. children who were short as a consequence of not producing enough of this hormone. Recombinant hGH could radically increase supply.

Multiple interpretations of the use of hGH are included in the current analysis of technological innovation processes because uses and markets ultimately work to select products and firms. Selection determines whether or not firms will recuperate their investments in innovative activities, make a profit, generate earnings to invest in new R & D activities, go bankrupt, etc.

By the early to mid-1980s, using genetic engineering to make human proteins in bacteria could no longer be termed science fiction, as it had been in the mid-1970s. This context of use had become industrial reality and soon marketable products. The emphasis in this chapter is on the pharmaceutical rather than the technology to produce it, although the two are closely related where changes in one often lead to changes in the other.

When initial production systems integrating genetic engineering had been designed and began functioning, an abundance of supply of hGH was possible. Moving from a situation of scarcity (pit hGH) to one of abundance (rDNA hGH) has been accompanied by negotiations over new uses of the final product, some medically legitimate and some illegitimate. There are also discussions to expand the definitions of the original disease, that is, dwarfism, to include short children within the normal height range. Negotiations over acceptable use are based on the results of medical knowledge-seeking activities about hGH.

Establishing and expanding multiple uses for hGH expands the market, giving firms incentives to fund this type of knowledge-seeking activity. Expanding the market was suddenly possible because genetic engineering would allow firms to obtain control over the source. Bacteria were amenable to control. In contrast, pit hGH faced a number of restrictions on expanding supply, such as the limited number of human pituitary glands available and the amount of hGH which could be extracted from each. With genetic engineering techniques, supply was only limited by uses defined by government regulatory agencies and by economic constraints, particularly how much a firm could sell at a feasible price for legitimate uses.

The first section examines how multiple interpretations of the use of rDNA hGH have been defined in a process of negotiation. A major theme of this section is how needs, uses, and markets were defined and sometimes redefined in relation to medical knowledge, experiments, and tests.

The next section examines a firm-level jump to a new rDNA hGH without the extra methionine (met-less hGH). This involved a new set of genetic engineering techniques, with related changes in production. Genentech developed a functioning technique to cut off the extra methionine in 1983. This change required that corporate managers and researchers make decisions about whether to finance new investments in new knowledge-seeking activities geared to the specifics of these genetic engineering techniques and biological materials. Otherwise they could not switch. It can be said to be a jump to a new trajectory, because of the many changes and basic parameters (model system) which also had to be changed. They had incentives to do so due to both scientific and economic arguments for the superiority of a metless hGH.

The succeeding section takes a look at negotiations over the definition of quality between rDNA and pit hGH. The latter was initially considered superior because it was natural. Comparisons and evaluation were based on medical scientific knowledge, which was one reason that Kabi's experience and knowledge with pit hGH was considered valuable. The superiority of the 'natural' version over the rDNA was, however, turned on its head in 1985 when doctors discovered a severe medical risk with pit hGH, leading to a ban in many countries.

In the countries in which it was banned, the market was suddenly wide open for the rDNA versions. The reactions of producer firms to these events is analysed in relation to their relative positions in developing genetic engineering techniques. This is thus another example of the social definition of selection criteria, here of the market as a whole.

Economic and technical aspects of innovative activities are dearly intertwined during different aspects of the innovation process analysed here. The relative success of Genentech and Kabi in relation to the size of the market is discussed at the end of this chapter, in order to indicate the relative success of their innovations.

MULTIPLE INTERPRETATIONS OF USE

Interpretative flexibility about how and why hGH could be used to treat different medical conditions has arisen from different understandings of what hGH can and cannot do in the body. Under conditions of limited supply, treating dwarfism was its only legitimate use.

Production of hGH based on genetic engineering techniques revolutionized this shortage. There was a great expansion in supply, leading in turn to controversies over appropriate uses. The controversies seem to be due to the fact that hGH is involved in metabolic processes of regulation in the body about which medical researchers have theories but do not completely understand. Some of its functions are known whereas others are still hypotheses. In short, the pituitary gland in the brain produces growth hormone, which in turn:

Stimulates the liver to produce somatomedins-hormones that are very similar in structure to insulin and that stimulate bones to grow. Children and adolescents produce large quantities of growth hormone, and adults normally produce very little of it.

But researchers believe that the hormone's primary purpose may not be to stimulate growth at all its principal function may be to conserve muscle tissue at the expense of fat tissue during times of stress. This journalistic summary indicates that medical researchers believe that hGH plays a role in various bodily functions other than growth, but that these roles are still unclear and in doubt. These roles are under negotiation and are hence uncertain.

Because growth hormone possibly has multiple functions in the body, it may be used to treat many other conditions than dwarfism caused by hGH deficiency. Other medical uses cited in the medical literature include retention of muscles in elderly people; as a dietary aid for obese people; to build muscle mass in athletes; as a way to make short 'but normal' children taller; and to treat 'burns or other catabolic conditions, steoporosis, peptic ulcer disease, or ageing'. In fact, some of these alternative uses of hGH were mentioned in Genentech's 1979 scientific paper using genetic engineering techniques, specifically the treatment of bone fractures, skin burns, and bleeding ulcers. When carrying out the initial scientific activities involving a genetically modified bacteria, Genentech and Kabi were thus aware that the potential markets for hGH were greater than the existing market for pit hGH (dwarfism).

How, then, should use of this pharmaceutical be understood? The term 'use' could be narrowly defined as patients' use of the pharmaceutical for legitimate purposes, that is, those purposes already approved by government regulatory agencies. This approach obscures, however, the social basis of these markets.

Legitimate uses are here seen as the result of negotiation processes in a locus of interaction between patients, doctors, firms, regulatory agencies, social conventions, and social protests. Negotiation over acceptable conditions is based on scientific research. This process is international because medical knowledge, like basic science, is international and because firms act to establish similar criteria in all their international markets. At the same time, the negotiation process is national because national regulatory agencies are institutions making somewhat different decisions and demanding slightly different tests. This includes establishing criteria to differentiate legitimate from illegitimate uses.

We can first look at the use of hGH to treat dwarfism and short children within the normal height range. When the only supply of hGH was extraction from human pituitary glands, the ever-short supply led to a rationing mentality among doctors and in definitions of the disease. Only those children shown to have unequivocal growth hormone deficiency received treatment. Unequivocal deficiency was fairly easy to spot by administering accepted tests.

In this situation, not giving the hormone to one child near the borderline meant that another needy child could receive the treatment. Short supply, leading to a rationing mentality and a strict definition of the disease, hence in turn influenced the definition of medical conditions that doctors believed desirable to treat. The most severe cases could be identified and treated, and questionable or uncertain ones ignored. As discussed earlier, the director of NPA argued in 1979 that 'all patients with diagnosed hyposomatotropism are receiving the hormone. "We have no waiting list."' From his perspective, there was no need to develop an alternative way of producing hGH because the strict definition of disease equated the number of potential patients to current patients.

Other medical researchers have argued that the supply of pit hGH only seemed adequate because the criteria for treatment were so very stringent. Michael Thorner of the University of Virginia Medical School argued in retrospect that when hGH was always in short supply, the idea was to be sure that 'every child who received growth hormone was growth-hormone-deficient, not that every child who was growth-hormone-deficient received treatment'.

The difference is quite important in relation to the size of the market. With the advent of plentiful supplies of hGH through genetic engineering-based production, all children who were growth-hormone-deficient could be treated, and supply could still exceed demand. Doctors were then given the real possibility of giving hGH to a larger number of children, including those near the borderline. As a result, doctors/medical researchers had to reconsider their criteria for deciding who should and should not be treated. Doctors were faced with the necessity of developing new, more sensitive medical tests to separate potential patients from non-patients.

The very medical knowledge about hGH thus came under negotiation in the early to mid-1980s when it was thought that supply would soon increase. Medical practitioners began to argue that 'normality' of hGH was not necessarily a static level which could be precisely measured but, instead, that normality included a range of conditions. In a popular account of this range:

Malnourished children] produce excessive amounts of growth hormone, yet they do not grow. Adults, who normally do not make measurable quantities of the hormone, begin producing much more of it when they fast or when they are stressed. Runners, for example, synthesize measurable amounts of growth hormone.

Under certain conditions, people who usually had normal levels produced more, and even within the category 'well-nourished children', no absolute line seemed to exist between normal and deficient. This fuzziness of definition of 'normal' offers possibilities for expanding areas of treatment. That line had to be agreed upon by medical practitioners in relation to social values of height. The medical scientific community could redefine standard levels of normal hormone secretion as well as what constituted a valid medical test. Firms, of course, funded doctors to do such research.

Doctors had existing standard tests to decide which children should be treated with pit hGH, but these criteria came increasingly into question when supply increased. This began when Kabi and Genentech were producing enough of their respective rDNA hGHs to have clinical trials. In November 1983, the National Institute of Child Health and Human Development held a conference on 'the uses and possible abuses of biosynthetic human growth hormone'.

The NIH and the community of medical researchers wanted to develop new, adequate criteria to decide whether or not to give treatment to patients on the borderline as well as to air some of the problems they had encountered, or felt likely they would encounter with increasing supply. When this conference was held, Kabi and Genentech did not have approval to sell their products, but they were in clinical testing. The conference attracted about fifty experts, from doctors to regulatory officials to corporate researchers. During the conference, the FDA wanted participants to try to come to agreement about medical uses of genetically engineered hGH and about appropriate testing procedures.

In fact, these conference participants disagreed about many things, indicating the fluidity of scientific knowledge. The discussions mainly centred around using hGH to treat pituitary dwarfism and borderline cases. In relation to that, they disagreed 'about which peak growth hormone values after pharmacologic stimuli, if any, constitute an adequate response. In other words, they did not agree what test results (peak values) indicated hormone deficiency. They similarly disagreed about which tests under what external conditions were optimal to diagnose the disease.

They disagreed whether and why some children seemed to have 'bioinactive growth hormone'. 'Such patients have growth retardation for which no cause can be found, apparently normal immunoreactive growth hormone secretion, and acceleration of linear growth when given growth hormone.' These patients and their reactions to hGH did not fit with the accepted paradigm of explaining children's failure to grow, and hence did not fit into the accepted idea of treatment. Because these cases did not fit, they indicated that currently accepted tests could not identify all cases where children could potentially benefit.

The disagreements clearly indicated that the potential demand to treat short children was much greater than the actual demand expressed under a situation of limited supply of pituitary hGH. It also indicated the fluidity of knowledge, here changing due to expanding possibilities of use and indicating the necessity of developing new test methods.

These conference participants did agree on two things. The first can be seen as recognition of their own ignorance, even when making medical decisions and diagnoses:

It was agreed, however, that the traditionally accepted discriminator of 7 to 8 ng per milliliter is arbitrary and that provocative tests are not always reliable for determining whether insufficient growth hormone is rate-limiting for growth.

The medical researchers thus admitted there was little correspondence between their standardized tests for diagnosis and occurrence of the disease. This uncertainty was partially related to the previous short supply, which curtailed the possibility of clinical trials to establish knowledge about such things. Before, even if they had established more sensitive tests to determine more precisely where the border lay, they would not have been so useful because they could not have had enough hormone to treat them. The context of potential use changed, putting pressure on specialists to develop new methods and to change the social criteria for selection of patients.

Secondly, the participants agreed that they needed to educate physicians and the public. They believed that spreading information was the only way to prevent unrealistic expectations about hGH. These unrealistic expectations could lead to severe disappointment and psychological problems in short children who did not respond to hGH treatment, and could fuel the black market trade for illegal uses. In short, expectations of miracles had to be tempered, and uncertainty in diagnosing and treating the diseases be admitted.

Firms have an interest in expanding current legitimate uses by questioning existing tests and standards and by clinical trials on new uses. Increasing the number of patients by redefining and expanding the initial disease would thereby increase demand. In addition to firms' actions, other individuals have placed expectations on the capability of hGH to do things for which it was not intended, that is, illegitimate uses.

This includes, for example, weight-lifters using hGH in an attempt to increase muscle mass. From the firms' perspective, more uses translate into larger demand, but at the same time, pharmaceutical firms have an interest in restricting sales to legitimate uses in order to protect their reputations and hence future sales. Firms, government regulatory agencies, and medical researchers thus have a common interest in defining criteria for legitimate uses and controlling illegitimate uses. They define selection criteria for legitimate uses.

The most controversial uses of hGH have included giving it to the elderly and to athletes and making 'normal but short' children taller. Much of the controversy over the use of hGH has centred on whether it is morally, ethically, clinically, and/or legally correct to treat these three conditions. There has been some public debate, in one case leading Genentech to temporarily break off a clinical test on short children within the normal range of the height curve.

The largest public controversies over hGH are rooted in philosophical questions concerning tampering with the human body. The basic dispute seems to be whether or not cosmetic changes or improvements can be considered real changes because they do not treat or cure the human body. Simply improving the body through pharmaceuticals is thought unacceptable. This notion of unacceptable tampering is so strongly rooted that it is embodied in law: pharmaceuticals must treat a disease not increase overall physical or psychological health. Neither ageing nor shortness in itself nor insufficient muscles are diseases, and therefore, if firms want to get these uses approved as legitimate, they would have to relate these conditions to a disease.

Medical uncertainty about testing procedures and about where to draw the line to define the disease has been important to the firms, because they could hopefully expand the medically accepted definition of the disease by sponsoring knowledge-seeking activities, clinical tests, conferences, and so forth. For example, similar questions to those mentioned above about appropriate measurements were addressed in a conference sponsored by Genentech in 1986. Obtaining adequate financial resources for such activities could also require creativity.

As an illustration, Genentech developed a limited clinical partnership to be able to afford clinical testing of hGH, namely the Genentech Clinical Partners, Ltd. Genentech essentially sold shares in the risk-taking of clinical trials in order to raise short-term money and spread risks. They later bought back the shares. Having the financial resources enabled Genentech to fund the generation of new knowledge and thereby change the interpretations of their product. In doing so, they could fund medical scientific research which would influence selection criteria defining market entry as set by regulatory agencies.

Pharmaceuticals differ from many other technologies in the high degree of negotiation over use between firms, regulatory agencies, and doctors/medical researchers. This social interaction defines the environment and selection criteria. They together define criteria for legitimate treatments based mainly on expanding medical knowledge. In the case of hGH the accepted criteria defining treatment have been increasingly called into question as external conditions change (increased supply). Moreover, when there has been little agreement over basic tests and standards, regulatory control has been weakened, especially over borderline cases.

Technical Implications

Parallel with the medical science discussions over use which started in the early to mid-1980s, managers and researchers at Kabi and Genentech might have imagined that they had completed the major part of their knowledge-seeking activities to design production systems. New challenges remained, however, when they were faced with the possibility of changing to a new set of genetic engineering techniques.

Researchers at the two firms had solved many of the challenges of developing new, commercially relevant production technologies integrated with genetic engineering. Kabi's and Genentech's technical achievements in designing, testing, and choosing a particular configuration of systems would allow the firms to manufacture hGH and compete on the market. The firms had also generated standard, more generalizable knowledge about commercial production of genetically modified bacteria, some of which was published in articles. This process is never finished, however, and the firms have continued to make many incremental improvements to the core and component technologies as long as they have been producing the product.

In 1983, however, Genentech researchers made an improvement to the genetic engineering techniques and biological material which required a larger change. In other words, Genentech and Kabi had to decide whether to continue making incremental improvements within their firm-specific trajectory, or whether to jump to a new set of genetic engineering techniques, following a somewhat different trajectory. Jumping would require investment in new knowledge-seeking activities.

On 25 April 1983, Genentech filed for an American patent for a new technique. In the new bacteria expression system, these researchers used a new signal sequence to direct *E. coli* cells to cut off the extra methionine on hGH by secreting it through the first cell membrane. This improvement enabled them to make hGH with 191 amino acids, like that produced in the body. Under the 1978 R & D contract, Kabi were entitled to any improvements which Genentech made, so both firms had access to these results. Producing hGH without the extra methionine was one of the challenges facing corporate use that had not been solved in the early 1980s.

In 1983 Kabi and Genentech thus faced a technical and economic choice. They had two alternative paths to follow, namely ignore the change or use it. A decision had to be made, based on a number of questions. Should they adopt this change in genetic engineering techniques? Was it an advance in technical and/or economic terms? What changes would it entail for the specific production system that each firm had already developed after significant investment in knowledge seeking activities? Would the improved product, the met-less hGH, give them an advantage on the market?

How much resources would have to be invested in new activities to solve the context dependent problems and specific design aspects? These were the general types of questions facing research managers in particular. They could, in fact, already identify dear technical and economic advantages in the new genetic engineering techniques and the 191-amino-acid hGH, but the firms had a difficult choice to make because Genentech and Kabi were already partially locked in to a technological trajectory based on the 192-amino-acid (met) version.

Both Kabi and Genentech had been working on the met hGH for four years when Genentech developed and gained ownership (patent) over this novel technique. The firms had thus invested considerable time and resources in the met hGH. They had invested within the firms to develop a functioning production system, and they had invested in collaboration with doctors and government regulatory agencies to negotiate legitimate uses through clinical testing.

Some of the knowledge, techniques, and equipment so developed could be transferred to the met-less version and would be directly useful, but some investment would be lost, worthless. The amount invested in met hGH but which could not be transferred would indicate the degree of positive reinforcement of existing choices about genetic engineering and biological materials. Such positive reinforcement was due to the accumulation of competences within firms, the regulated nature of entry into the market, and the related, systemic technical choices. The question was whether Kabi and Genentech wanted to and could jump to a related but different trajectory for metless hGH.

In fact, Genentech and Kabi felt they could not just drop the met hGH version until they had developed a functioning production system for the met-less hGH. This might take a couple of years, and during that time, they needed to be able to sell the met hGH. If they chose to develop the met-less variety, each firm would thus have to develop two alternative systems and clinical trials in parallel. In making this choice, one of the main questions that both firms had to answer in 1983 was, is there room and resources within the firm to develop the met-less version? Can we afford diversity? Each would have to bear the costs of investing resources in knowledge-seeking activities to develop that diversity.

According to persons interviewed, some individuals in each firm thought it would be better to concentrate all the resources on one project, that is, to concentrate on getting the met hGH out on the market. This corporate strategy placed priority on early entrance into the market by concentrating resources. It would select less diversity. Others thought it would be better to divide the firms' resources into two projects because the new met-less alternative was superior. More specifically, if they did not change, any other firm making 191-amino-acid hGH would in the future claim that their version was superior to Kabi's and Genentech's products. It might prove to be so clinically as well, thereby marshalling scientific research to support that claim. That would erode Kabi's and Genentech's future markets.

Erosion of market shares was particularly threatening because hGH is a long-term. proposition. Children use the medicine for about ten years, and doctors only change brands if there are major problems or major improvements. This long-term lock-in of doctors and patients to the product of a particular firm thus gives advantages to those firms which can quickly move into the market with the highest quality and 'closest to nature' product.

Within Kabi, the project leader Linda Fryklund strongly argued that Kabi ought to pursue the met-less version in parallel, even though the met hGH had been brought so close to market. She was convinced the change was necessary to compete in the market. Within Kabi, resources were made available for more R & D, and deadlines set. This indicates that research management had to judge the extent to which apparently technical changes also entail economic considerations. Genentech was initially more ambivalent but put some resources into the project. Both firms were quite reticent about discussing their decisions publicly, trying to keep their new projects a secret.

Both Genentech and Kabi thus decided to work on the two hGH projects in parallel, and initially worked independently. As indicated above, jumping to a new set of genetic engineering techniques and biological material would require that the firms invest in knowledgeseeking activities to design a new production system. Even though some of the knowledge, techniques, and equipment previously generated, selected, and retained within the firms could be transmitted, the corporate researchers would also have to generate and select new solutions to specific problems. Knowledge and solutions were related to general categories of opportunities and challenges but would again have to be modified for context-dependent use.

One of the first challenges encountered was that fermentation was more difficult. The new bacteria host was more finicky about growing inside the fermenters. As Florell of Kabi put it:

As an aside, I can say that it was really lucky that we started with the first bacteria strain [met] instead of the second met-less.

The met version was easier from the cultivation point of view. It was easier to reproduce. It didn't go wrong as often. I think that the very hGH gene part is more stable. The non-met version was more sensitive; the parameters must be exactly right in certain phases.

The parameters of fermentation thus had to be more carefully controlled because the met-less hGH was more sensitive and hence went 'wrong' more often. The corporate research groups, however, soon developed specific biological conditions for fermentation which were better suited to these finicky bacteria. Precise control and testing were needed to do so. Such difficulties with the fermentation process made these steps more expensive than for met hGH.

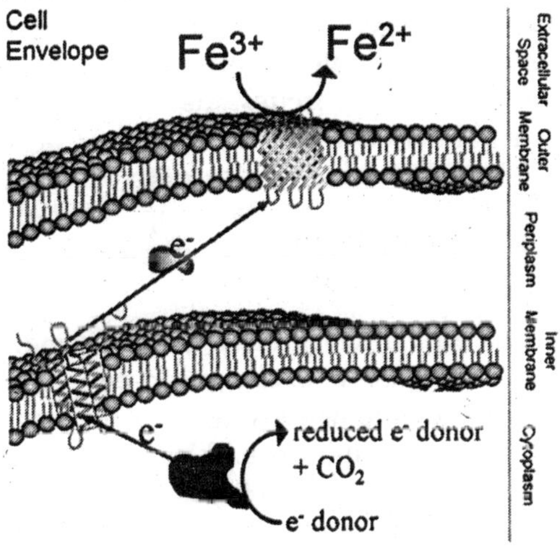

Fig. periplasm

The protein purification steps also had to be changed, but these steps could be simplified. Instead of producing the protein within the cytoplasm of the *E. coli* bacteria, Genentech's new technique secreted the hGH in between the two membranes of the cell, into the periplasm. Secretion cut off the signal sequence from hGH and thereby got rid of the extra methionine.

This hGH was also soluble and folded correctly. Many of the initial challenges to purification thus disappeared with this major change in the genetic engineering material. Although the bacteria also secreted many other proteins into this periplasm, the firms would no longer have to tear the cells apart to get at the desired protein. A simple water based solution could extract the hGH. The consequence of these changes was that the number of purification steps could be reduced, and so these steps were less difficult in many senses. In contrast to increased cost of fermentation, this worked to reduce costs of production. These two examples indicate the integrated nature of the production process.

On the one hand, Genentech and Kabi benefited from having first developed the met hGH production system. They benefited in that they knew some specifications of engineering and scientific techniques — such as specific parameters and analytical methods — that might be useful for the new system. The firms had already generated, tested, and selected solutions for previous specific problems, but these often represented categories of problems. Discussions among managers and researchers established in-house technical and economic criteria to help select among alternatives. They had also established some common criteria for approval of genetically engineered pharmaceuticals through discussions with various regulatory agencies.

Genentech only interacted with the American FDA, whereas Kabi interacted with European and Asian agencies as well; the pattern of interactions followed the firms' projected market areas. In short, the firms had learned from being early movers into recombinant hGH and could transfer some experience, competence, and knowledge to the new project. On the other hand, the general pattern of technological innovation repeated itself with the met-less hGH. New technical challenges were identified, and solutions suggested by the corporate researchers at Genentech and Kabi. Some challenges were specific; others general.

These challenges were commercial as well as technical in those different configurations of the production system affected cost of production, and ultimately, selling price and profits. Moreover, having decided to support diversity by developing the two hGHs in parallel, the Kabi R & D group in particular were under enormous pressure to work rapidly to bring the new product to market.

Genentech and Kabi initially worked separately, but in February 1984, Genentech called Kabi and wanted to collaborate. Kabi were interested in co-operation partly because it would speed up their project, and time of entry into the market was essential to them. Moreover, a new person was in charge of the rDNA hGH project at Genentech, William Bennett, and Kabi employees found it easier to communicate with him than had previously been the case.

When Kabi and Genentech started to collaborate in 1984, they were potential rather than actual competitors. They had divided up the rDNA hGH world markets, with Genentech retaining rights to the American and Canadian markets. Kabi intended, however, to continue selling their pit hGH in the United States even after Genentech began selling its rDNA version, making them potential competitors. They had no direct competition over recombinant DNA products, so the two firms had some incentives to co-operate if each could benefit.

They co-operated loosely rather than engaging in a truly joint development project, where each firm would have performed specific subprojects under a unified goal. Their collaboration took more the form of exchange of information about problems, possible solutions, dead end solutions, and so forth. They met once or so annually and had weekly telephone discussions, usually a handful of people crowded around a speaker telephone on each side of the Atlantic discussing problems or ideas. Before the telephone discussions, data was sometimes sent by fax, an emerging technology.

The reason this collaboration was important was that it gave each specialist group someone else with whom to discuss the directions and results of their knowledge-seeking activities. They could discuss new challenges, how to identify problems, what solutions worked and what did not work, try out new ideas which were a bit off the wall, and so forth. In doing so, they would jointly identify some of the contours of their common search space. Communication between the corporate researchers enabled each group to narrow down the broad set of all possible choices to a few alternatives which seemed most likely to function technically at a reasonable price. Firms needed to reduce diversity.

Hearing what had and had not worked for others was one way to reduce diversity and fill in the blank spaces on their respective maps without having to do everything themselves. Thus, because of the trial-and-error nature of this type of engineering work, where theoretical models could not replace experimentation in contexts of use, learning from each other's empirical mistakes and successes could accelerate the pace of technical developments.

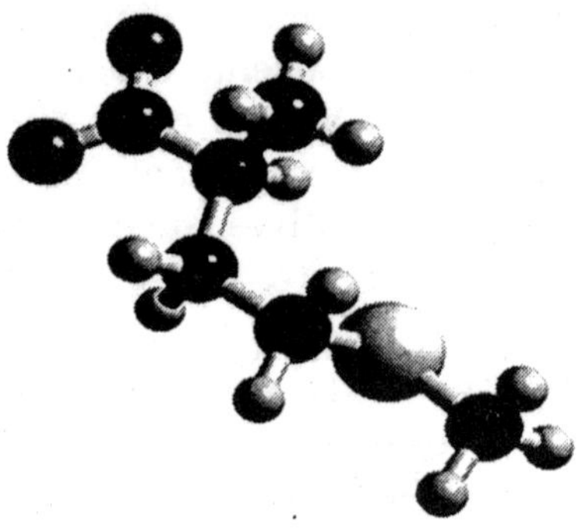

Fig. Methionine

Communication with the larger community of researchers was largely cut off during this initial R & D phase. Neither firm wanted to disclose publicly that they were working on this new project, or the technical problems encountered. They were keeping their knowledge-seeking activities fairly quiet, even within the broader population of engineers and scientists potentially interested in these questions. Genentech had put in a patent application for the technique to cut off the methionine, indicating to whomever was interested that they had solved that part of the problem, but the firms neither confirmed nor denied rumours about the project. Collaboration between these two firms partially replaced contacts with the larger community.

This meant that contacts with the other firm were the main outside source of ideas and information during certain phases of technical development. Occasionally, university researchers working on processoriented problems were included in development work but kept under oath of secrecy. For example, Kabi's process R & D group did not want to interact with university researchers in Sweden because of the risk of leakage of information. They were trying to prevent competitors from imitating their techniques and hence being 'free riders'. Instead of risking leaks, Kabi hired a researcher from the Department of Biochemistry and Biochemical Technology at the nearby Royal Institute of Technology on a limited time contract. She worked for a few years on Kabi's hGH production system, under oath of secrecy, before returning to the university.

Kabi found secrecy to be the best way to safeguard details of the process. Genentech was similarly secretive. The generation of new knowledge about specific technical details and even about the existence of major improvements was seen as giving competitive advantage.

The firms did not want to divulge details such as the exact parameters of optimization or the exact order of steps. The value of this knowledge lay in the specifics of translating knowledge into use, which generated economic benefits and entrepreneurial profits. Other knowledge was by its nature more general and often public, and could be distributed through published papers or generalizable problems and solutions taught in textbooks.

The firms could contribute to that body of knowledge and still retain secrecy about details. For example, the firms sometimes published four or five different ways of accomplishing the same production engineering task, thereby not divulging which was most effective, or which the firm actually used.

In developing their two relatively similar production systems for metless hGH, Genentech and Kabi engaged in a process of developing relevant knowledge and techniques. As with the earlier met hGH, search and innovative activities required many employees' specialized knowledge as well as interactions with agents in other organizations.

On the one hand, these relations with others were important sources of information and were part of the social process to define selection criteria. On the other hand, many of the specifics of context-dependent knowledge and techniques were generated or known by employees. Most of the equipment, the physical manifestation of production, was rooted inside the firms.

Shaking up hGH Markets

Despite Swanson's earliest plans to try to get Genentech's recombinant hGH approved by 1982, January 1985 had come and gone without any national regulatory agencies approving either Genentech or Kabi's hGHs. In 1984-5, Kabi were hoping for Swedish approval in 1986 and Genentech for American approval somewhat earlier. The firms were close to commercialization but had faced dead ends and challenges delaying approval, although clinical trials had also been positive. Once approved, they would face competition from pit hGH, which was often distributed by government agencies free of charge. In addition to price, they had to sell a product based on a radically new technology in a market with a 'natural' alternative.

Interestingly for the firms and their perceptions of markets, the acceptability of genetically engineered hGH was being defined in relation to the human-produced pit hGH. Defining a genetically engineered pharmaceutical product as legitimate was as important as defining legitimate uses for hGH as a pharmaceutical (see second section). Like hGH in general, defining legitimate uses for rDNA versions was a process of negotiation between medical researchers, firms, and government agencies, based on medical knowledge.

The firms thus had incentives to promote the superiority of genetic engineering techniques in these negotiations. The two firms differed a bit. As seen by its very name, Genentech – GENetic ENgineering TECHnology – was proud of its genetic engineering proficiency. No one in the firm needed convincing of the superiority of the new technology.

Kabi was a more biological pharmaceutical firm, sensitive to public debate, and less concentrated on the technology *per se*. Kabi was careful in naming its first rDNA hGH to avoid mention of genetic engineering and stress instead its everydayness, namely 'Somatonorm'. Interestingly enough, the name of Kabi's second, met-less, hGH proudly displayed its genetic engineering origins; the brand name is 'Genotropin'. These names were directed towards external agents, such as doctors and patients, and reflected the firms' interpretations of their demands.

Even within Kabi, some employees initially viewed genetically engineered products with suspicion. Many thought that hGH extracted from pituitary glands was the natural, the real stuff and that the rDNA produced version would be second-rate. Marketing people within the company even complained that they would not be able to sell it. Although scientists including the project leader believed in rDNA hGH, others argued that there would be little or no potential demand because a 'natural' alternative existed. Management in the firms also knew, however, that potential demand was greater than existing supply of pit hGH.

The perceived relative superiority of rDNA versus pit hGH changed quite dramatically on 2 April 1985. This changed the market as well. 2 April was the day that the American Food and Drug Administration (FDA) announced that they had found one positive case of a very rare viral infection of the brain called Creuztfeldt-Jakob disease. The person had been treated with pit hGH supplied by NIH in the late 1960s and early 1970s.

The disease was carried by a 'slow virus', an infective agent which could lie dormant for up to twenty years, and it was thought that contaminated glands had introduced the virus into the hGH solution. This former patient died. A death involving pit hGH thus prompted government regulatory agencies to downgrade radically its perceived quality. This correspondingly opened the market for alternatives. It should be noted that the contaminants in rDNA hGH were marginal compared to this.

The first indication of trouble had come in March 1985 when a Stanford University physician informed the NIH that an autopsy showed that this young man had the disease. The research organization (and research financier) NIH turned the problem over to the government regulatory agency, namely the FDA. The NIH also immediately halted their clinical tests of pituitary hGH which did not involve therapy.

The FDA acted to control this unwanted and dangerous side-effect. It made an announcement about the one positive case on 2 April and then sent corporate employees and NIH and university researchers home to see what could be done to ensure safety — particularly how to test for purity, and how to destroy slow viruses — and set another meeting for a few weeks later. On 19 April the FDA announced to the firms that three more cases had been discovered, although one diagnosis was uncertain.

Based on this knowledge, they asked commercial distributors in the USA, which were Kabi and Serono Labs, to withdraw pituitary hGH from the market, but made it dear that if the firms did not, the FDA would. The NIH similarly stopped production.

Later in April, the FDA announced that a new case had been discovered. This time it was in the United Kingdom, with hGH supplied by the Medical Research Council. As this contaminated hGH had come from a somewhat different extraction process, it indicated that slow viruses could be a general problem for tissue extraction. Soon after the FDA took action, similar bans were imposed in Sweden and in a number of European countries, excluding France. The natural hGH from pituitary glands could no longer be argued to be superior to the genetically engineered one, which could not be contaminated with slow viruses.

Kabi prepared to take its brand name Crescormon off all their international markets and made its announcement on 24 April. Not everyone, however, agreed that pituitary hGH should be pulled off the market just because a few former patients had contracted and died from CreuztfeldtJakob disease. The medical knowledge and actions of the FDA could be, and were, questioned by agents in both the techno-economic and basic scientific environments. These bans were important for our tale because they closed the market to all firms who had not switched to genetic engineering as the basis of production.

There were many uncertain factors about this decision, one of the most important of which was the purity of the pit hGH that these former patients had received compared to contemporary standards. Before 1977, the pit hGH provided by the NIH was between 25-50 per cent pure, meaning that the remaining 50-75 per cent of the preparation were unknown proteins and contaminants. This was what the patients contracting Creuztfeldt-Jakob disease had received. In contrast, some argued that the pituitary hGH available in 1985 was up to 95 per cent pure.

No evidence could be shown to support the notion that the purification techniques available in 1985 posed a danger of contamination by slow viruses, but the problem was that neither could tests prove that no danger remained. For example, a UCSF medical researcher specializing in this type of virus argued that the NIH had no choice but to halt distribution of pit hGH. In the same article, he argued that there was no evidence that contemporary hGH was contaminated. This time, it was the safety of the pharmaceutical from 'natural sources' like human pituitary glands that was under scrutiny and debate.

The positions and arguments of firms on the question of whether or not it was appropriate to ban pit hGH can be related to their contemporary knowledge-seeking actions to implement genetic engineering. Because Genentech did not supply pit hGH and was one of two firms – along with Eli Lilly – poised to move into the American market with rDNA hGH, a ban on the pituitary version could only be beneficial to them.

In fact, immediately after the FDA's withdrawal of pituitary hGH, the FDA gave Genentech an investigative new drug status 'for the time being' for the 150 children who could die without it. However, the majority of children took hGH for dwarfism and could forego treatment for a maximum of six months without sabotaging their final height. They did not immediately get Genentech's product. These six months would, however, be a tight deadline, putting pressure on the FDA to act quickly.

In addition, there were two firms supplying pit hGH to the commercial segment of the American market, for children who had not been receiving it from the NIH. These two firms were Kabi and Serono Laboratories (Massachusetts), a subsidiary of Aros-Serono (Geneva). Both companies sold hGH internationally, and Kabi had about 20 per cent of the American market whereas Serono had about 50 per cent. The two firms reacted oppositely to the FDA's announcement and subsequent ban.

When Kabi had gone home after the first FDA meeting on 2 April to evaluate the different possibilities of testing and destroying slow viruses, they knew that they had an alternative in the form of their rDNA hGH, brand name Somatonorm. Kabi quickly came to the conclusion that they could reduce the probability that pit hGH was contaminated with viruses infected with Creutzfeldt-Jakob disease. However, they could never conclusively say that purification eliminated all such viruses. No test could be definite, and so there would always be a risk involved.

Moreover, Kabi thought the necessary changes to reduce the probability of contamination were expensive. Each batch of hGH would have to be tested, and the virus destroyed in two separate ways. Secondly, destroying the virus in two separate ways might destroy some of the hGH, thereby reducing yields. Both of these changes would be costly. In a much later account, Fryklund motivates Kabi's decision with the idea that 'what was unethical in the US was unethical for the rest of the world. Bodies work the same.' Given an alternative at hand, the pit hGH brand name 'Crescormon' was not worth saving.

Nevertheless, Kabi did have to make a difficult decision in that they could have continued selling the pit hGH in a number of countries. It was difficult in that they had to stop distribution immediately and buy back the supplies stockpiled in subsidiaries around the world, ultimately leading to a loss of about 150 million Swedish crowns ($21.4 million dollars). The company as a whole then went from a profit to a loss because Crescormon was their largest product. Even this option was costly, at least in the short run.

In contrast to Kabi, Serono had moved into genetic engineering for hGH much later (starting in 1984) and were not far along with developing a production system, much less clinical testing. Serono Laboratories made the opposite decision about the American market and did not stop distributing its pit hGH until 9 May 1985, when the FDA made a formal request.

Serono argued that its purification methods removed all viruses and felt they should be allowed to continue selling it in the USA and internationally. Serono's director Francesco Pilato maintained that the 'FDA has no evidence that Serono's product is not safe.' A Serono spokesman similarly said, 'We think the federal government] acted much too hastily' and also expressed doubts about the very diagnoses of Creutzfeldt-Jakob disease in the patients.

It was a difficult diagnosis to make, and perhaps the doctors were wrong. Although Serono employees undoubtedly believed their own interpretations of facts about the safety of their testing methods and product, the fact remains that the firm had no viable alternative. Serono continued selling pit hGH in those countries where it had not been banned, as did, for example, the French firm Sanofi in France.

Serono had only moved into genetic engineering to produce hGH a year earlier. This can be compared with Kabi's six years. In January 1984, Serono signed an agreement with the British biotech company Celltech for an expression system. They were hoping to start clinical tests in late 1985 but were having some technical troubles and challenges. New challenges had arisen because they had chosen to use mammalian cells — rather than bacteria — to produce hGH. Mammalian cells grow much more slowly but were thought to construct the protein more faithfully. In short, Serono's rDNA hGH was considerably further from the market than those of Kabi and Genentech, and Serono thought that their pituitary hGH was worth the cost of saving. The FDA, however, forced them off the American market, as did regulatory agencies in a number of countries afterward.

At the end of October 1985, the FDA approved Genentech's product, called Protropin, for the American market to treat certain kinds of dwarfism. It took almost exactly six months after the first FDA meeting, and six months was the limit for breaking off treatment before it affected the children's final height. Approval took only twenty-three months, about half of the normal approval time.

In connection with approval, the FDA demanded that Genentech design a 'market survey system' in order to gather information about each patient, including, for example, growth rates and side-effects. The FDA saw this as a way to maintain close control. Genentech originally saw this system as an expensive extra burden, but it turned out to be an effective marketing device for convincing doctors of the efficiency and safety of their product. A large body of medical knowledge about the positive effects on other patients was persuasive as a selling argument.

When FDA approval came, Genentech was ready with an offensive marketing strategy. According to an article in *Business Week*, Genentech had been stockpiling the product for almost a year and had plenty to start shipping. Genentech had also developed plans to build a top sales force.

Genentech systematically snared top members of the hospital-based forces of such drug companies as Merck, Pfizer, Upjohn, and Eli Lilly, hiring them on a contingency basis. When FDA approval seemed imminent last month, Genentech called in the chits.

Boasts President G. Kirk Raab, a 25-year veteran of Abbott Laboratories who joined Genentech last February: 'We've got not just top sales people but elite groups who've been working in academic hospitals and have long-term experience dealing with physicians.'

Genentech wanted to move quickly into the market. They knew that even if the hGH market was wide open with the removal of pit hGH, their product would not completely sell itself. They knew that competitors would soon want to enter the large American market, and there were advantages to being first and capturing market share. In a strategy similar to their hiring experienced production process people, Genentech consciously went after salespersons with relevant experience.

In contrast to Genentech, Kabi had experienced salespersons in-house, who already had established contacts with academic hospitals and specialized physicians. The salespersons were now convinced of the superiority of rDNA hGH. In the longer run, however, Kabi's former marketing strategy based on demand's exceeding supply would have to change to meet the new condition of multiple producers, multiple uses, and the possibility of supply's exceeding demand.

Kabi had expected approval later, not sooner, and had a large international market to cover. When Kabi got approval in their home Swedish market in October 1985 and soon in other countries, they did not have a large enough manufacturing capacity within Sweden to meet demand. They had to find an alternative production site. Kabi once again turned to the government-owned Porton Down in England. After the original R & D contract of December 1980 had run out, Porton Down had run fermentations a second time from 1983 to 1984.

This was part of Kabi's R & D work to design a production process, and part of making enough hGH for toxicology and clinical trials. In 1985 Porton Down hoped to become a licensed producer for Kabi, using Kabi's technical specifications, and the two signed a five year contract for manufacturing. Even before that contract lapsed in 1990, however, Kabi had decided that Porton Down's production was too small and too costly. After this contract lapsed, this small percentage of production was discontinued, and all production returned to Sweden. In the mean time, Kabi scaled up its factory outside of Stockholm, leading to new technical challenges to be solved.

The removal of pit hGH from the American and many European markets in 1985 thus changed attitudes about the desirability and safety of genetically engineered pharmaceuticals. Even 'natural' alternatives extracted from human bodies could be contaminated, and in fact, the genetically engineered version tested much purer. The six-month deadline put pressure on government regulatory agencies to accept the new product. The medical knowledge generated through firms' clinical trials thereby created and legitimized the market, after negotiations over criteria. This negotiation process was about the product as a pharmaceutical, including the strict criteria for safety. Genetic engineering offered an alternative pharmaceutical product which appeared safer than the natural version in 1985, leading to new interpretations of uses and markets.

Monopoly and Competition

In most countries, competition among firms over hGH increased as new entrants into the market began producing in the 1980s and 1990s. The largest single market, the American one, is somewhat different. In 1983, the US Congress passed a law giving firms making 'Orphan Drugs' a seven-year monopoly. The purpose of this federal legislation was to encourage firms to produce pharmaceuticals for small patient groups. The argument was that without exclusive market and monopoly rights, other firms could copy the idea and enter the market.

This 'free ride' would thereby reduce incentives for the initial firm to innovate because with that risk of imitation, no one would be willing to invest the substantial R & D expenses necessary to develop a new drug for a small and potentially unprofitable market. With a limited-time monopoly, firms would be encouraged to take risks because the chances of recuperating costs were higher. Genentech received an 'Orphan Drug Status', giving them a seven-year monopoly on the American market along with FDA approval in October 1985.

Although Genentech got a monopoly, the FDA simultaneously expressed a desire for competitors in the market. Dr John Gueriguian, who had called the FDA conference on insulin and human growth hormone made with recombinant DNA back in 1981, clearly stated their position when interviewed by *Chemical Weekly* in 1985:

FDA's Gueriguian insists, though, that the agency has not simply handed over the whole market to Genentech. FDA, he says, has been 'extremely active' in encouraging 'numerous' companies to pursue development of synthetic HGH during the past four years.

The fact that FDA expressly stated its interest in competition may help explain another twist in the American hGH market that happened in 1987. Genentech had received approval in 1985, but in March 1987 the FDA approved Eli Lilly's genetically engineered hGH for sale. This was thus during Genentech's seven-year monopoly. Moreover, the FDA also gave it Orphan Drug status. What was the difference between the two products that permitted both Orphan Drug status? The difference was the one amino acid, the methionine, still on Genentech's brand name Protropin. Eli Lilly used a bacteria expression system essentially identical to Genentech's, but then cut off the methionine using enzymes. The FDA therefore ruled that Lilly made a different product.

The law on Orphan Drugs was written fairly vaguely, and there was uncertainty both about how the FDA would rule beforehand and why they did so afterward. Genentech was caught by surprise and thought the decision was a violation of the Act and threatened to sue the FDA. This decision seemed particularly unfair to the Genentech researchers, because they had in fact already developed a new production system for a met-less hGH and carried out much of the clinical testing. They were so close. The FDA's decision would exclude Genentech from the met-less hGH market until 1994, although they could and would continue to sell their met hGH. After the Orphan Drug status expired, Genentech replaced Protropin (met hGH) with Nutropin (met-less) in the American market, and both Kabi and Genentech are considering moving into each other's market in the future, when the exclusive market contract expires.

Before the FDA approved Eli Lilly's met-less hGH, Genentech knew they must have been dose to submitting material. Genentech knew about their clinical testing, and Kabi had told them around March/ April 1987 that Eli Lilly had applied for approval in Sweden, to challenge Kabi's potential monopoly position. Both Lilly and Kabi were approved in Sweden in the spring of 1987. Lilly also had strong incentives to get American approval. They needed American approval because home country approval is necessary to sell internationally, even if it is an informal requirement for doing so. Serono Labs were also close to submitting an application for met-less hGH in the USA. Kabi had urged Genentech to act quickly to get the met-less hGH approved in the USA.

Genentech instead took some extra time, which delayed their FDA application.

They decided to make a number of improvements which would make production more efficient but which took time to implement. For example, they further revised the bacteria to increase yield, which thereby reduced the cost of the product but also necessitated changes in related production technologies. Kabi turned down some of these improvements in order to get approval more quickly. Genentech were also slow about sending in the enormous amount of documentation required for new drug approval.

Genentech managers felt that time was less urgent than Kabi had suggested, partly because they did not think Eli Lilly had done enough clinical tests with met-less hGH and partly because they never thought the FDA would approve a second hGH. Genentech lost out in that Eli Lilly's got Orphan Drug status. In effect, Genentech and Lilly both got monopoly rights to the same market. The one-amino-acid difference was clinically unimportant for growth.

Genentech have not, however, lost out on sales. The American market was worth around $200 million in 1991 and about $300 million in 1995. Of this, Genentech has had 75 per cent of the market. Genentech sold about $157 million worth of Protropin in 1991; $205 million in 1992; and $216 million of Protropin/Nutropin in 1995. In 1992, Eli Lilly sold $170 of hGH internationally. Thus, despite the fact that Genentech were cut off from developing the improved version, Genentech have been more successful in the American market than Lilly.

The other international markets are more competitive. hGH has been a commercial success as a product and for some early-moving firms. The world-wide market for hGH in 1991 was estimated at $700 million; at $1.96 billion in 1992; and around $1 billion in 1995. There has been an enormous increase in sales since genetic engineering has made increased supply possible. Kabi (subsequently Pharmacia, then UpJohn Pharmacia) have also been successful and hold about 55-60 per cent of the market outside the USA. Kabi sold $243 million (1.7 billion Swedish crowns) in 1992, of which 90 per cent was exported from Sweden. Genentech's first two R & D projects using genetic engineering to make known pharmaceuticals — insulin and human growth hormone — have thus turned out to be enormously successful.

Both the American and international markets have thus turned out to be several times larger than the firms had originally projected in 1978, with the American market soon worth dose to $400 million rather than $10 million and the international market worth around $1 billion. The original calculations of sales, markets, and profits were thus way off due to the expandable nature of multiple uses and markets for this product. The early-moving firms have been rewarded by profits, although Genentech has been rewarded more than Eli Lilly. The firms expect increased competition and lower prices in the future.

This chapter has concentrated on knowledge-seeking activities directly affecting or creating uses of, and hence markets for, hGH. Much of the focus has been on how firms identified and used technical and economic criteria to direct their activities to conform to existing, and create new, social selection criteria. For pharmaceuticals, selection criteria to enter the market are set by government agencies, based on medical research. Although pharmaceutical markets are created and controlled by government regulatory agencies, their decisions are based on knowledge created in interaction and negotiation between firms, government agencies, and medical researchers, which creates conditions for entry as well as the market. By defining legitimate uses, those uses excluded from the market are illegitimate uses, often obtained through the black market.

The firms had incentives to invest in knowledge-seeking activities to generate legitimate uses of the pharmaceutical. In the best case, medical research would show multiple uses (indications) or else expand the range of uses for existing indications. If approved, these would increase the market for rDNA hGH, which was possible in a situation of expanding supply. In fact, moving from a situation of very restricted to expanding supply has led to new discussions about where boundaries for treatment should lie.

Another important selection criterion was homogeneity and high quality of the product. The firms needed to convince the involved parties that both the product and its production system were standardized, reproducible, and conformed to regulations through more sensitive analytical methods. The firms' generation of novel, legitimate uses of rDNA hGH were thus dependent on their ability to generate medical scientific knowledge and to influence selection criteria defining the market.

In 1983, Genentech and Kabi had to decide whether or not to develop a second version of rDNA hGH, a met-less hGH that was identical to the hGH produced in the human body. Partly based on others' basic scientific research, Genentech modified the general knowledge and techniques to develop this specific technique. Their technique cut off the signal sequence containing the methionine and secreted the protein into the periplasm. Deciding whether or not to change to this new set of genetic engineering techniques was a fairly difficult decision.

Genentech and Kabi had already sunk investments into the met hGH, in the form of knowledge-seeking activities, designing production systems, clinical trials, relations with government regulatory agencies, and so forth. Changing to a new set of genetic engineering alternatives would significantly improve the perceived characteristics of the final product but would also entail significant additional R & D activities. A decision had to be made by management.

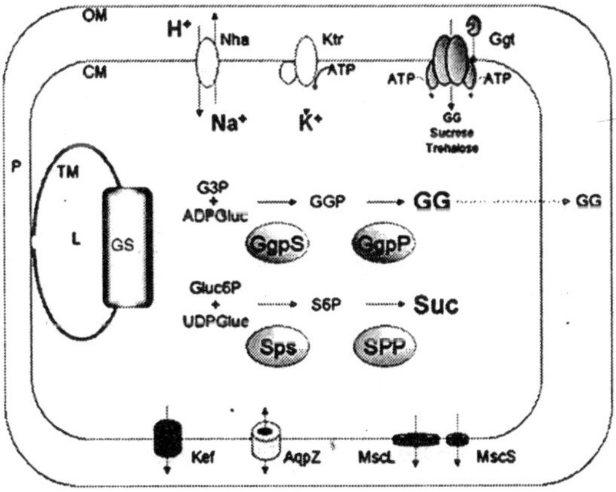

Fig. periplasm

The sunken investments could have led to a firm-level lock-in to the met hGH. Both firms decided, however, to work on the two projects in parallel. Having made the decision to sustain diversity by developing both versions, Genentech and Kabi knew they ultimately wanted to stop producing the met hGH, and so they had to jump to the met-less trajectory. Initially, they had to work on producing both versions because the new version would take several years of R & D and clinical testing to be ready for sales. They made this decision because in the longer term, the firms wanted a high-quality met-less hGH in order to expand market share. Whether the met-less hGH was actually better clinically or not, its structure was identical to the pituitary hGH, making it more attractive to doctors and patients.

Genentech and Kabi firms initially worked independently, but soon co-operated in exchanging information in order to speed up the innovation process. It gave each group someone to discuss opportunities and problems with in a situation where the firm did not want to divulge information, not even to researchers responding to the basic scientific environment. Information could easily move from those researchers to competing firms.

When Kabi and Genentech developed their respective rDNA hGHs, they assumed it would compete with pituitary hGH, as sold by firms or provided by government sponsored research institutes. Many initially argued that the pituitary hGH was superior. It was 'natural' whereas the rDNA was artificial, produced by a radically new technology. Ability to increase supply with the new technology was not a sufficient marketing argument for those children lucky enough to be receiving pit hGH, although purity could be a good argument. The relative superiority of -the two was quickly revised, however, during the mid-1980s. A fatal medical risk with pit hGH was uncovered in 1985, leading the United States, Sweden, and many other countries to ban pit hGH. This left the hGH market wide open for those producing with genetic engineering techniques.

The reactions of firms to these bans can be related to their relative positions in developing genetic engineering techniques for production. Genentech was positive and able to step immediately into the American market to supply the neediest children. This was the first pharmaceutical that they produced by themselves, and their sales of hGH rocketed to $41 million in 1986. Kabi was in a more complicated situation. Pit hGH was already an important product, and taking their brand name Crescormon off all. their international markets meant a big financial loss for the firm.

At the same time, doing so would be best for the children and favourable for their reputation (an important aspect of the pharmaceutical business), and they had an alternative product in waiting based on a different, low-risk technology. Even for those countries where it was not banned, Kabi destroyed their share of the market by stopping Crescormon sales, hoping, of course, that they would soon fill the vacuum with the rDNA version.

Other firms which had previously made pit hGH but which had not moved into genetic engineering contested the dangers of slow viruses when modern purification techniques were used, and continued selling when and where possible. In contrast, firms previously not selling hGH but interested in entry saw the switch to genetic engineering techniques as an opportunity to move into the market. Government regulatory agencies thus destroyed the market for pit hGH by banning it in many countries, thereby changing the market and putting pressure on themselves to approve shortly thereafter a genetically engineered version (within six months). Associating severe risks with the pituitary hGH dramatically increased the relative superiority of the rDNA hGH.

The new met-less hGH was ready for sale in the mid- to late 1980s. Kabi were particularly successful in quickly jumping to the new trajectory and developing 'Genotropin' to sell internationally, whereas Genentech were somewhat slower, and the FDA shut them out of the metless segment of the American market in 1987.

Government regulatory agencies could encourage competition or else monopoly over the pharmaceutical. The US Congress had designed a law granting a limited-time monopoly for new pharmaceuticals intended for a small number of patients in order to encourage innovative activities.

Genentech was granted such a monopoly on sales of met hGH in 1985. In order to increase competition anyway, the FDA approved the met-less hGH of Genentech's competitor Eli Lilly in 1987.

This effectively shut Genentech out of that segment of the market until 1994, but Genentech has successfully competed with their met hGH, then their met-less or natural hGH. They have about 75 per cent of the American market and will consider moving into the international market after the agreement with Kabi expires. Kabi has also been successful internationally, with about 55-60 per cent of the world market, excluding the American market. Competition was thus not stopped with the ban of pit hGH in 1985, but competition was thereafter over the rDNA hGH of different producer firms in most countries.

The generation and selection of uses and markets for pharmaceutical hGH has depended on negotiations and interactions particularly among producer firms, medical researchers, and government regulatory agencies. Their interpretations of use and markets have set social selection criteria, defining conditions of entry into the market.

Chapter 4

Genetically Engineered Microorganism

FROSTBAN

Roger Smith of Monterey, California, answered his front door early one evening in February 1986 and found June Wilson from the house around the corner with a petition to force the county Board of Supervisors to prevent the field testing of something called Frostban right in Monterey. She said that many environmentalists were worried that Frostban, which was a bacterium engineered using the new molecular genetic techniques, could be dangerous to people, plants and the environment. Some said it could be a severe environmental threat, even possibly stopping the rain.

Frostban was developed by a company called Advanced Genetic Science (AGS) as a way to protect plants, including strawberry plants, from frost damage. And the Board of Supervisors was worried that if something did happen as a result of the release to harm someone or cause damage to plants or animals that the town would be sued.

There was no insurance for something like this, and the town could lose a lot of money. Mr. Smith was part owner of the local John Deere store, and he knew firsthand from the strawberry farmers who came into the store that they could lose a whole year's berry crop to a bad frost. If they didn't make money, then they didn't have money to spend on equipment, and his business might be in trouble. So why should he be against just testing this product in a small field? June Wilson told him that Frostban was unnatural, a bacteria made in the lab, and that no one really knew what would happen if it got into the environment. It was wrong to release an unnatural bug, and it might be dangerous to them, their kids, their gardens, who knows what. The company and the scientists didn't have the right to experiment on them.

What should he do? He worried about his family, and the environment, but how dangerous could this be? Wasn't a permit required from the state or the federal government? The EPA was always making inquiries about his insecticides. They were pretty careful. What did they say about this? June said the EPA scientists had said it was OK, after reviewing the proposal from AGS, but there were still many other scientists from California, Washington and even from Europe who said that not enough was known to be sure that it was safe to release the bacteria into the environment. She said that you couldn't trust the company and the university scientists. Hadn't one of them been fined by the EPA for testing that very bacteria on a roof-garden in Oakland last year?

How should Mr. Smith decide whether or not to sign the petition?

The bacterium that AGS wanted to test was a genetically engineered strain of *Pseudomonas syringae* a bacterium commonly found on plants. The genetically mutated version was intended to slow or prevent the formation of ice crystals on the plant when the temperature dropped below freezing. In the 1970s scientists had learned that two bacteria found on plants, *Pseudomonas fluorescens* and *Pseudomonas syringae,* have on their surface a protein that serves as a centre for ice crystal formation. Ice will form only if three conditions are met: water is present, the temperature is sufficiently low and there is a molecule to catalyze the formation of crystals of water, the transition of water from a liquid to a solid—thus ice. It is the formation of ice that causes the damage to plants when the temperature falls below freezing.

As water freezes it expands, mechanically destroying the cells and structures of the plant causing it to die and allowing spoilage, the bacterial breakdown of the plant material. Without a crystallizing centre, water can fall to nearly "40°C ("40°F) without freezing, without solidifying. Keeping the air temperature above freezing around the plants, with smudge pots or fans, is the conventional approach to preventing frost damage of vulnerable crops such as strawberries or oranges. But it is an expensive process because workers have to start and tend the pots, and it doesn't always work.

Fig. Pseudomonas syringae

In the 1960s researchers working on a corn fungus had discovered that a powder from infected leaves when dusted on plants in a greenhouse allowed the plants to freeze if the temperature dropped. Plants not dusted did not freeze, that is, ice crystals did not form. In the 1970s Steve Lindow, then a graduate student at the University of Wisconsin in Madison, deter-mined that it was a bacterium, *Pseudomonas syringae* in the leaf powder, not the fungus that allowed ice crystal formation. Ice would only form if a bacterial strain expressing a particular protein was present.

Such strains were called, ice positive, or ice nucleation positive (INA+). INA+ *P. syringae* was common on the plant; substitute INA" bacteria and the plants were able to tolerate low temperatures because no ice formed. Lindow and Deane Arny filed a patent claiming a method for reducing the temperature at which freezing takes place in plants by the addition of nonice nucleating bacteria. Later at Berkeley Lindow attempted to develop ice minus *P. syringae* by treating INA" bacteria with a chemical mutagen, but it was difficult to get a bacterium with only the critical gene mutated, without so altering the bacteria that it couldn't successfully compete with the wild type bacteria on the plant.

Using the then newly developed tools of molecular biology, Lindow, Orser and Papopolous replaced the functioning INA gene with a nonfunctioning gene, thus converting INA+ *P. syringae* into INA". It was the type of "molecular surgery" that rDNA technology was all about.

Only one gene was changed. The gene encoding the protein responsible for ice nucleation was replaced with a gene that could not express the protein. Advanced Genetic Sciences, a biotechnology company founded in 1979, had been working on methods for protection of crops from frost damage and had licensed the Lindow technology from the University of California (UC) which held the patent.

Testing within the confines of a laboratory greenhouse by the UC Berkeley group, supported the idea that if the INA" *P. syringae* were sprayed on plants, the plants could better tolerate low temperature, without freezing. But the lab was not the real world and if this approach was going to work, it had to work in the field. Specifically, field tests must show that if the ice minus bacteria were sprayed on plants, the bacteria would survive in sufficient number, not be eliminated by competition by the bacteria already on the plant and protect the plants from natural frosts.

But the NIH, the federal agency regulating recombinant DNA research, had strict guidelines prohibiting the release of genetically engineered organisms into the environment without review and approval by the director of the NIH, with advice of the Recombinant DNA Advisory Committee (RAC), a committee formed by the NIH to help it review the Guidelines for rDNA work. In 1982 Lindow and colleagues at UC Berkeley and AGS petitioned the RAC for approval for field tests of its INA" bacteria.

On June 1, 1983, the NIH authorized the release for field testing of the ice minus bacteria. However, Jeremy Rifkin, an author and head of the Foundation for Economic Trends, filed suit for an injunction prohibiting the experimental release on the grounds that the NIH had failed to meet the legal requirements of the National Environmental Policy Act passed in 1969. In May 1984 Federal Judge John Sirica granted the injunction, stating that the NIH had failed to meet the requirements for an Environmental Impact Statement. Subsequently the NIH worked with the EPA to review the petition from Berkeley and a similar one from AGS.

Releases were planned for AGS in Monterey and UC Berkeley at the Tulelake Agricultural Research Station. Local citizens continued, with the help of Rifkin, the Foundation on Economic Trends (FET) and other state, national and international environmental groups to block the releases.

The federal court's decision to grant the injunction in 1983 was based on a specific point of law: that the alteration of the NIH Guidelines in 1978, which allowed exceptions to the prohibition of certain experiments—including the release into the environment of a genetically engineered organism, provided these experiments were expressly approved by the Director of the NIH, with advice of the Recombinant DNA Advisory Committee—constituted a major federal action significantly affecting the quality of the human environment.

This meant that the requirement of the National Environment Policy Act (NEPA) for an environmental impact statement (EIS) applied. The judge also held that the NIH and the RAC did not have the information to make the determination. The plaintiffs had argued that neither the NIH nor the RAC had sufficient expertise to prepare the EIS, and thus the EPA was required by law to prepare the EIS. To prepare the EIS the EPA considered four items or issues:

1. The genetically engineered organism, including the parent organism, the sequences added or deleted and the methods used ;
2. The planned release, including the plants and animals that could be compromised by the release ;

3. The potential for the ice minus bacterium to establish itself in the environment, based on information on its survival, multiplication, dissemination and transfer;
4. The effect of the ice minus bacteria on the environment

One trigger for the alarm was a worst case scenario for the release of the ice minus bacteria published by a highly respected environmentalist. He wrote that rainfall could be reduced by the establishment of the ice minus bacteria if the wild type ice plus nucleation protein was a factor in the ice formation in the atmosphere necessary for rainfall. AGS held that the assumptions in this scenario were unproven and that the concern would be answered by the data indicating that the ice minus bacteria had no competitive advantage over the wild type ice plus bacteria in the environment and thus would not replace them.

Studies by the petitioners demonstrated to the satisfaction of the EPA that the parental *P. syringea* was not dangerous to humans, animals or any crops in northern California. Tests on the pathogenicity of the parental line and the ice minus bacteria were run on seventy-five different plants. The applicants also showed through a series of studies on greenhouse potato plants that the ice minus strains have no competitive advantage. In the middle of the review process and hearings on legislation about the regulation of release of the products of biotechnology into the environment, there was a flurry of negative publicity that put AGS, the EPA and others in an unfavorable light.

A story ran in the February 26, 1986, *Washington Post* revealing that AGS had tested Frostban out in the open. The EPA indicated that they were unaware of this release and publicly expressed much consternation. The facts were revealed in hearings on March 4, 1986, held by the Oversight Subcommittee of the House Science and Technology Committee, at which scientists from AGS and the EPA as well as public officials and private citizens from Monterey testified.

AGS had carried out studies of the pathogenicity, the ability to cause disease and damage, to a wide variety of plants, including some trees, as part of the safety studies required by the EPA. The studies on several types of trees were performed by injecting the genetically engineered bacteria, as well as control non-engineered bacteria and known pathogenic bacteria, into the trees.

What had been revealed by what the Post called a "source close to the study" was that the tree pathogenicity studies were done on the rooftop of the AGS building in Oakland. The Board of Supervisors of Monterey, because they learned about the exact location of the proposed field test from other sources, but had not been told about it by the EPA, held a public hearing on January 27 and then passed an ordinance banning for fourty-five days any release of a genetically engineered microorganism. Leon Panetta, the congressman from that district, came to the congressional hearings and expressed outrage that a field study was proposed to take place within a more or less residential area, less than ten miles from 10,000 people.

Sam Karas, the chairman of the Monterey Board of Supervisors, was particularly angry because he learned about the rooftop test from the *Post* story, after a visit to AGS when he had concluded that the AGS scientists were honest, careful and trustworthy. Since they hadn't told him about the unofficial release, the inference was that they were not to be trusted.

Also Karas and others were worried about their liability if something untoward happened as a result of the field test. Two alarms raised by Rifkin and his legal colleagues were;

1. That there was insufficient information to provide an accurate assessment of the risk of the release of a genetically engineered organism;
2. That there was no insurance protection from financial liability if something went wrong. Karas noted that AGS was a small company with limited resources, but the county was seen as having deep pockets. The chair of the subcommittee questioned the AGS representative about the plan to clothe the AGS field test technicians

and EPA technician observers with protective suits. Wouldn't that add to the fears of local residents? It was a concern, AGS agreed, but the EPA had insisted on such garb. In the end, AGS was fined on technical grounds related to failing to disclose fully the details of the conditions of safety tests in the application for the field test. And the publicity in Monterey, the ill will and the petitions of the Monterey citizens made AGS change their proposed site for the field test to an isolated spot in Brentwood.

The EPA went ahead and considered the release applications from AGS for Brentwood and from UC Berkeley for the Tulelake Agricultural Research Station. The applications and the safety data were reviewed not only by EPA scientists, but also by scientists at the U.S. Department of Agriculture, the FDA, the NIH. The EPA convened an advisory panel with two microbiologists, a meteorologist, a community ecologist, a microbial ecologist and a plant pathologist to provide its scientific review.

The EPA decided, with agreement from all the advisory groups, that the application of the genetically engineered INA- to 0.2 acres of field presented "no forseeable significant risk to human heath or the environment."

The first releases were performed in Brentwood, California, on April 24, 1987, and on April 20 at Tulelake Agricultural Research Station with EPA scientists in space suits supervising the release. The results of tests over four years and ten frosts (temperatures below 32°F but above 25°F) indicated that indeed the ice minus bacteria would survive and would reduce ice damage by 80 percent. However, this success did not lead to commercialization of genetically engineered ice minus bacteria in large part because of the significant negative publicity, the certainty that the approval process for commercialization would be long and expensive and that Rifkin and the FET would continue to raise legal challenges to commercialization.

In fact, AGS merged with another company, DNA Plant Technology (DNAP), in 1989. DNAP decided that the legal challenges raised by Rifkin would be more expensive than the sale of the genetically engineered bacteria would bring in profits and stopped research on ice minus in 1990. They developed and marketed a naturally occurring *P. syringae*, then licensed the technology to Ecogen, a small biopesticide company, in 1992. A company called Plant Health Sciences currently markets a naturally occurring ice minus bacteria named Blightban A506.

This case is significant despite the failure of a commercial product to appear because it was the first approved release of a genetically engineered microorganism into the environment. It defined the contending forces in the area, and it led to the establishment of the current process by which such releases are regulated in the United States. It also contributed to the establishment, through the Coordinated Framework for Regulation of Biotechnology published in the *Federal Register* in 1986, of the division of responsibility among the NIH, the EPA, the USDA and the FDA for regulation of biotechnology.

In each case, products are subject to pre-marketing approval. Under the Toxic Substances Control Act (TSCA), pre-market notification (PMN) had to be filed for use of microbial products of biotechnology. From 1987 through 1995 thirty-four PMNs for recombinant microorganisms were filed.

Fourteen of the PMNs were for environmental use, predominately for field testing of new strains of nitrogen fixing bacteria, to improve the yield of plants such as alfalfa. These were research tests to evaluate, using antibiotic resistance, the competitiveness of the new strains and their ability to enhance nitrogen fixation. One PMN was filed for a microorganism to be used for bioremediation, to break down polycyclic hydrocarbons in the soil.

The organism was a strain of *Pseudomonas fluorescens*, which had been engineered to contain a gene for naphthalene degradation from another *Pseudomonas* and a bioluminescent reporter gene from another genera. In 1994, the EPA proposed rules for regulating microbial products of biotechnology under TSCA and after a long period for comment and review, the final rules were published in the *Federal Register* in April 1997.

The reporting rules continued to be applied to "new" microorganisms, those organisms formed by combining genetic material from organisms in different genera, a classification based on the relatedness of the organisms. The new rules established two levels of reporting:

1. Research and development (R&D) for commercial purposes, by or for a commercial entity;
2. For commercialization, with greater flexibility and a shorter review period for the R&D.

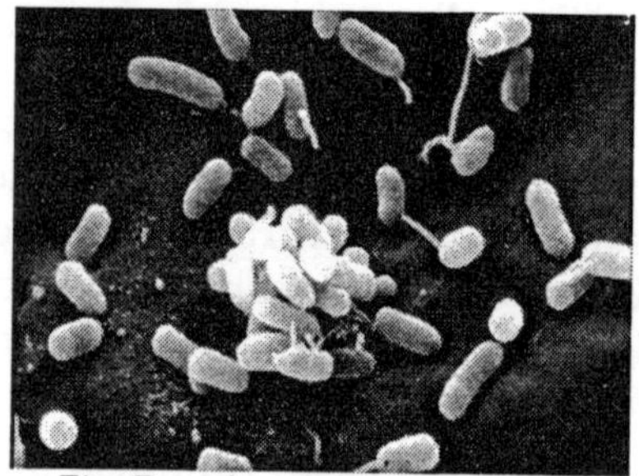
Fig. Pseudomonas fluorescens

Researchers at institutions that are in mandatory compliance with the NIH Guidelines for Research Involving Recombinant DNA Molecules, that is, that are receiving federal funds, are exempt from EPA reporting requirements. Researchers in voluntary compliance with NIH Guidelines can, by documenting their use of those Guidelines, satisfy EPA's requirement for testing in contained structures. Research involving release into the environment even by those otherwise following the NIH Guidelines, which prohibit release into the environment, will clearly not be exempt.

The requirement for premarketing approval and concerns about biotechnology has meant that products are judged not simply for what they are, but by how they are produced. As more experience has been gained with rDNA and biotechnology, there has been a slow, sometimes very slow, evolution by the regulators to address biotechnology products for what they are and not get mired in the process by which they are produced. This regulatory maturity comes too late for the genetically engineered ice minus bacteria. While other developments in biotechnology contributed to the formation of the current regulatory bureaucratic structure and to the activists' approach, ice minus was the starting shot.

Chapter 5

Genetically Engineered with an Insecticide

CORN

Mr. John Leary plants over one-third of his 400acre farm with corn. One of the problems he faces is the reduction in yield caused by the European corn borer (ECB). He deals with the borer by monitoring his fields, and if he finds infestation he applies chemical insecticides or the powder of the bacterium *Bacillus thuringiensis* (Bt), which produces a toxin. The problem is that both the chemical and bacterial treatments wash off, have to be reapplied later in the season for the second stage of the life cycle and are ineffective for the larval stages that are inside the plant. Recently several seed companies, including DeKalb, Monsanto, Ciba and Mycogen, have created corn plants that produce a Bt toxin active against the European corn borer (ECB). The yield of corn from Mr. Leary's farm fell by 10 percent in 1994 due in large part to the corn borer.

While he attempts to use chemicals sparingly, employs sustainable agricultural techniques such as the use of Bt spray and does not till over half of his farm to avoid erosion, he believes that corn borer insects survive over the winter in his nontilled fields and reduce the yield throughout by infesting all of his fields. He has tried several of the conventionally bred resistant hybrids in the past, but the cost was high and yields were not. The transgenic Bt corn seems like something he could try in part of his acreage this year since he read in the newspaper that Bt corn seed had been approved for sale by the USDA and the EPA. The salesman from DeKalb is offering a pretty good deal on Bt seed, but some of Mr. Leary's neighbours are concerned that planting it will lead to development of insects resistant to the Bt spray that they rely on.

The scientists cannot agree on the risk of resistance one way or the other. His son, who attends the University of Iowa said that a speaker at the University recently talked about concerns raised by the Union of Concerned Scientists about resistance. The seed salesman says that DeKalb and the other companies selling Bt have developed resistance management programmes, required by the USDA in its ruling, to allow the release of the genetically engineered corn.

Should Mr. Leary plant Bt corn? Agriculture is big business. The business of providing food and other agricultural products to the over 5 billion people in the world is a vast undertaking, but in much of the developed world it has become the work of the few. In the United States fewer than 2 percent of the population farm to provide food for the other 98 percent. To provide this efficiency farming must be highly productive and cost effective. Small decreases in yields or small increases in costs can turn a successful year into a losing year.

Corn is the largest single crop in the United States; over 78 million acres were devoted to corn in 1996, representing 25 percent of crop acres harvested. The corn harvest has a $30 billion value, based on a harvest of 9 billion bushels at $2.25 to $2.85 per bushel. This translates into 236 million metric tons of corn produced in the United States, over 40 percent of worldwide production. China is the next largest producer, accounting for 20 percent of world production. In the United States nonalcoholic beverages and foods account for over 45 percent of corn usage, alcoholic beverages 6 percent and fuel alcohol 24 percent. Corn is also used in a number of industrial processes.

A farmer grows and harvests his or her corn and sells it to wholesalers for a price that can vary year to year. Profits per bushel are small and variations in the price per bushel and/or small variations in the input costs to produce that bushel can make a significant difference for the farmer.

The seed corn that the farmer plants are purchased hybrids developed by the seed companies. The use of such purchased hybrid corn seed grew in the United States this century as the concepts of genetics were applied to developing stronger, better-yielding corn plants. The seed companies develop highly inbred lines of corn, each with certain desirable characteristics and then with controlled pollination cross two inbred lines to create a hybrid. The hybrids may demonstrate what is known as hybrid vigour, showing the best of both parental lines and then some.

Since hybrids do not breed true, that is, when the kernels of a hybrid corn plant are planted the resulting plant will have only some of the good traits of each parental line and not every trait of the hybrid. The conversion to hybrid seeds has resulted in a six fold increase of corn yield since the 1920s. Farmers thus buy new seed every year to achieve high yield. In fact, some seed companies require farmers to sign an agreement not to practice brown bagging that is, collecting kernels from the mature hybrid plants and planting them the next year.

The European corn borer is a significant pest of field corn in the Corn Belt; losses in yields and expense for control cost U.S. farmers more than $1 billion each year. The European corn borer, which falls in the order Lepidoptera, was accidentally brought to North America in corn imported from Hungary and Italy for broom making in the early 1900s.

Since its introduction it has spread to all but the most western states and changed from one generation per year to two, three and four generation populations in all but the most northern of affected states. The moth lays its eggs and the resulting larva bore into the growing corn plant and depending on the site of the damage and the maturity of the plant, the corn plant is shorter with fewer and smaller leaves. This reduces the availability of water and nutrients to the plant and may cause an inability to hold the maturing ear of corn—the stalk collapses.

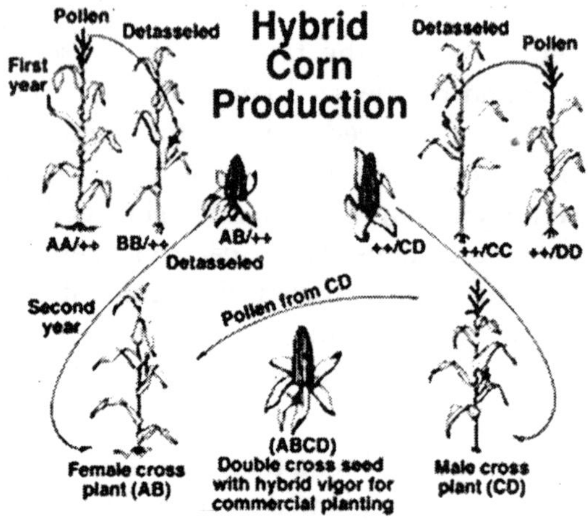

Fig. Hybrid corn plant

Yield loses from the ECB in terms of bushels of corn per acre can range from 2 percent to 33 percent. Seed companies have developed hybrids that show resistance to the first generation larvae, resulting in less physiological damage, but these hybrids have not shown overall high yields. Other hybrids compensate for increased susceptibility to second and third generation with better overall yield.

Control of the European core borer requires monitoring the field for evidence of infestation and treatment with insecticides. The decision to spray for the ECB is carefully made, based on the level of infestation found by a procedure called scouting. The farmer or scout picks at random a certain number of new leaves from plants and counts the number of corn borer larvae found. This allows him to estimate the average number of corn borers per plant.

Using the expected yield of corn from experience with the hybrid in use, the weather and location, the estimated market value of the corn and the costs of the insecticide and its application, the farmer decides whether the application of the insecticide makes economic sense. Since a number of components of the calculation are estimates, read educated guesses, the farmer could easily make a mistake deciding to treat or not to treat with insecticides.

The chemical insecticides used for the European corn borer fall into several chemical classes including organophosphates, carbamates and pyrethroids. They generally kill by affecting the insects' nervous system particularly by inhibiting enzymes that control nerve conduction. The labeling, use and restrictions on chemical insecticides are regulated by the EPA based on their risk to humans, animals, plants and the environment. The insecticides used for the corn borer generally are classified by the EPA as Restricted Use Pesticides, purchased and used only by certified applicators, because of their adverse effects on humans, birds or fish. Their application is associated with risk to the individual applying the chemical and the environment.

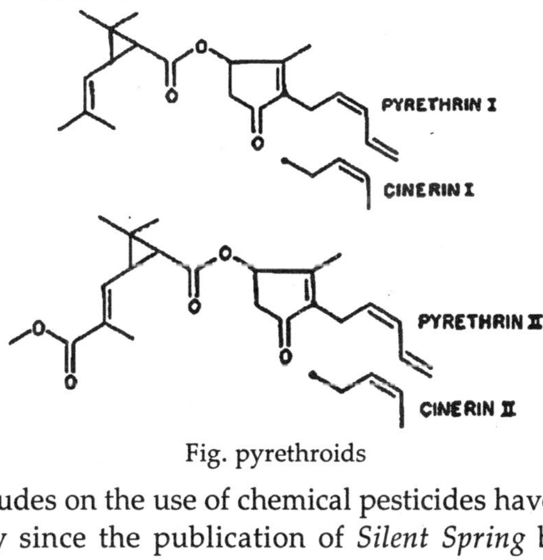

Fig. pyrethroids

Attitudes on the use of chemical pesticides have changed markedly since the publication of *Silent Spring* by Rachel Carson in 1962. The environmental movement sparked by the publication of that alarming book led to passage of the National Environmental Policy Act of 1969. NEPA is the basic national charter for protection of the environment. It establishes policy, sets goals, and provides means for carrying out the policy.

Its purpose is to instruct federal agencies on what they must do to comply with the procedures and achieve the goals of the act. In 1970 the Environmental Protection Agency was formed to implement the goals of NEPA, in large part by integrating and extending activities that had been the responsibility of other federal agencies. In particular, the EPA took over the role of regulating pesticides from the Agriculture Research Services of the Agriculture Department.

Concerns for the environment, the safety of farm workers and consumers and the need for soil conservation have supported the development of a movement to what is termed sustain-able agriculture, which is directed to balancing the environ-mental, social and safety issues without significant long-term economic harm to the farmer.

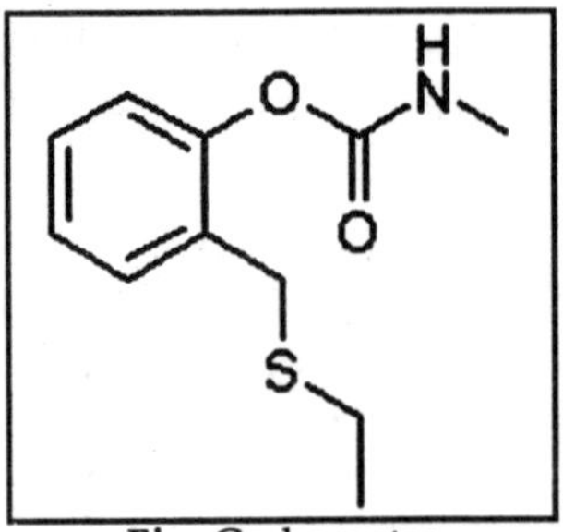

Fig. Carbamates

One expression of the move to a more sustainable approach to agriculture is a more prudent use of chemical insecticides and an interest in what are termed biopesticides, the use of natural biological approaches to the control of pests. The use of *Bacillus thuringiensis* is a good example of a biopesticide. It has been known since 1911 that the bacterium *Bacillus thuringiensis*, a common soil organism, killed the larvae of certain moths. It also became clear that different strains of *B. thuringiensis* kill different insects. In each case Bt kills insects with alkaline digestive systems through the action of a crystalline protein toxin accumulated in the spores of the bacteria and ingested by the insect. The toxins are referred to as Cry proteins and more than fifty have been isolated and identified. They fall into six families based on their selectivity, the specific insects that they kill.

The Cry protein when ingested into the alkaline gut of the insect dissolves and then is activated by gut enzymes to a form that attaches to receptors on the gut wall. The selectivity of each Cry toxin for particular insects is based on the receptor/ Cry protein match. Once bound to the receptor, the Cry protein forms a pore in the wall of the gut, which disrupts the ability to exclude water and retain certain ions. The insect dies because of the destruction of the gut leading to massive tissue water uptake.

Fig. Bacillus thuringiensis

Bt as dried bacteria was registered with the Department of Agriculture and approved as an insecticide in 1961. It has subsequently been approved by the EPA. It is safe for humans and the natural enemies of many crop pests. Safety studies in humans indicated that humans could eat a gram of Bt per day for five days with no medical complaints. Bt was one of the first biological control agents registered for use against insects in the United States. Safety data submitted by the manufacturers indicate that Bt did not have acute toxicity on birds, dogs, guinea pigs, mice, rats or humans. Bt of various strains are used to control spruce budworms, gypsy moths, cabbage worms in broccoli and cauliflower, budworms in cotton and tobacco and leaf rollers in fruit.

The World Health Organization (WHO) has used Bt in West Africa to control blackflies. However, not much Bt is used by corn farmers. Less than 1 percent of pesticides used in the

United States contain Bt. In 1995 less than 0.4 percent of corn acreage in Iowa was treated with Bt; 2.6 percent of acreage was treated for the European corn borer. The disadvantages of Bt for the farmer are its relative low activity, its vulnerability to sunlight and the fact that rain washes it away from the plant and the insect. So it must be applied at the right time and place and may require repeat applications. This makes the calculation for the cost effectiveness of treatment even more uncertain and risky.

Plant seed developers have made a business, in fact an industry, of improving the germplasm of crop plants through conventional breeding techniques. The corn seed industry is based on the development of highly inbred parental strains each with some desirable traits and the generation of hybrids that express the best of both parents for one generation. The investments of the companies in the development of the hybrids are protected by biology and by law. The hybrids, as noted, do not breed true and thus the farmer must, to get the best yield, buy new seed each year rather than collecting seed from his harvest.

Patent law protects an inventor by prohibiting anyone else from selling the invention for a period of time after the patent was filed, if the patent issues. Such laws protect the seed companies' investment in developing the hybrids by prohibiting anyone from selling the specific hybrid for eighteen years after the patent is issued.

The intellectual property rights of the horticulture and seed industry have been given patent protection in the United States through the passage of legislation separate from the Patent Act. As the plant industry was developing in the 1920s as a result of the introduction of the new genetic insight, companies sought protection for their investments in the traditional way, by the availability of protection of what is termed intellectual property, through patents that give the patent holder exclusive rights to sell the invention for seventeen years after the patent issued.

However, there was resistance to patenting plants under the Patent Act in part because of a widely held belief that plants were products of nature and as such not patentable. The success of Luther Burbank and other plant breeders in asexual propagation of new horticulture varieties was a strong impetus for passage of the legislation. In fact, Burbank's widow sent a letter in support of the bill and the secretary of Agriculture testified on its behalf. Congress and the administration saw the horticulture business as an important new business in need of patent protection to motivate investors and inventors. In 1930 the Plant Patent Act was enacted providing patent protection only for plants produced asexually because there was no confidence at that time that plants produced sexually, from seeds, bred true.

By the late 1960s further work on plant breeding, primarily at state agricultural experiment stations, had demonstrated that sexually reproducing plants could breed true to type. Legislation extending patent protection of the PPA to sexually propagated plants was sought. The driving force behind this legislation was private industry seeking protection of financial investment in the research and development of new sexually produced plants and feeling a growing concern for a threat to the dominance of American companies in the agriculture business coming from Europe. In 1961 the International Union for Protection of New Varieties of Plants (UPOV) was formed in Europe to protect national breeders' rights, resulting in anxiety that U.S. breeders would be at a disadvantage in international markets.

Efforts to amend the PPA failed in 1968; new legislation was sought, and in 1970 the Plant Variety Protection Act became law providing protection from competition for eighteen years for inventors of a new variety of sexually propagated plant. Two exclusions to that protection are of interest:

1. A breeder cannot exclude others from using the protected variety for research.
2. Farmers have the right to save a seed/crop for their own use and the use of other farmers. The latter exclusion is

of limited relevance for corn farmers since hybrid corn does not breed true, and seed saved and used the next year is unlikely to provide the same high yield.

It was a natural extension of the investment in producing better corn germplasm using conventional breeding techniques to seek to apply the advances in the genetic engineering of microorganisms. Plants were relatively attractive as raw material for genetic engineering particularly because the ability to clone a plant from a single cell was a technique well known since the 1920s. In this process adult plant tissue is separated into single cells and with the application of the right cocktail of plant growth hormones, each single cell could develop into an entire plant. Several companies decided to combine the genetic engineering technology with the ability to clone plants to make transgenic plants.

Corn—because of the experience in its genetics, the reliance of the farmers on purchased seed, the size of the market and its narrow profit margin—seemed a good place to start. The Bt Cry genes were a natural for introducing into plants because of familiarity with the toxin and the availability of the toxin genes. The advantages of the Bt biopesticide, particularly its safety record, was a significant factor especially if the transgene were to be introduced into a food crop.

The major disadvantages of the Bt biopesticide, that it was destroyed by sunlight and washed off by rain and was ineffective against late stage larvae inside the plant tissue, also added to the argument in favour of a genetic engineering approach since the Bt transgenic plant would have the toxin in place, unaffected by sun and rain, able to kill the corn borers throughout the season without requiring reapplication.

The raw materials to make a transgenic organism were all in place: a gene that would add value to the hybrid seed and a process to provide the cell, which, after genetic modification by the new technologies, would allow development of a whole plant. What was needed was a way to get the desired transgene the DNA construct, into the plant tissue culture cell and a way to select, from all the cells in culture, which ones had been successfully genetically engineered. For microorganisms the delivery vehicle for the gene to be introduced is a bacterial virus called a plasmid, which will infect the bacterium and result in stable incorporation of the DNA construct either as an independent genetic element reproduced along with the bacterial chromosome upon cell division or as an insert into the bacterial DNA. For some plants there is a bacterium that carries a plasmid that will do the job.

This bacterium, *Agrobacterium tumefaciens,* lives in the soil and infects certain plants. Upon infection the *Agrobacterium tumefaciens* releases a Ti (tumor-inducing) plasmid that causes a tumor to grow on the plant. The tumor develops because the Ti plasmid inserts some of its genetic material into the genetic material of the plant, leading to the production of proteins and nutrients needed by the *A. tumefaciens* and causing loss of growth regulation of some of the plants cells. The resulting mass of bacteria and plant cells develops into a tumor called a crown gall.

To genetically engineer plants susceptible to *A. tumefaciens,* molecular biologists removed from the Ti plasmid the genes causing the tumors and inserted the gene or genes of interest, put the altered plasmid back into the bacteria and then infected plant cells with the genetically altered *A. tumefaciens.*

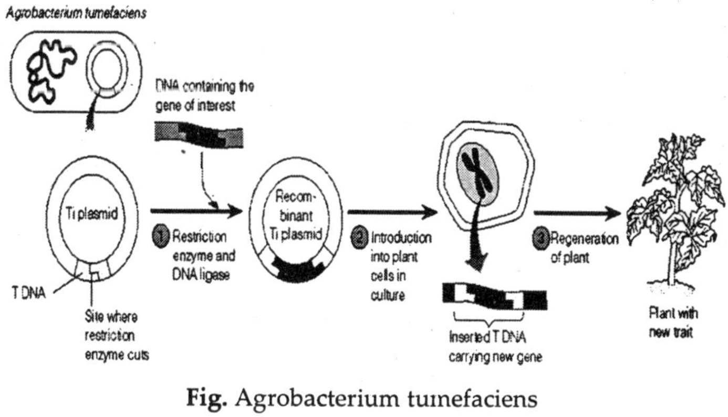

Fig. Agrobacterium tumefaciens

Corn is not susceptible to *A. tumefaciens*, so other methods had to be used to deliver the DNA to the tissue culture plant cell. These are physical or chemical methods, which generally are used on plant cells that have been treated with enzymes to break down the carbohydrate cell wall, leaving the cell membrane behind. The resulting cell is called a protoplast. The DNA construct can be injected into the protoplast with microsyringes made from capillary glass. This is very laborious and since the introduced DNA integrates into the plant cell DNA only rarely, many cells have to be injected.

A large number of protoplast can be made leaky with the use of chemicals such as polyethylene glycol or by exposure to an electric field, a process called electroporation. Another technique that has been developed to introduce DNA into cells involves shooting the DNA into the cell, a process called ballistic or biolistic.

The DNA is coated onto tungsten spheres and shot into the cells with compressed gas. An improved version of this technique employs a brief pulse electric field to propel gold particles coated with DNA into the plant cells. This form of the biolistic technique for genetic engineering of plants is the most advanced and the most successful in getting the DNA into the plant cell with the least chance of damage.

Once the DNA construct is introduced into the plant cells, the next challenge is to determine and isolate those very few cells in which the DNA has been integrated into the plant DNA in a way that allows the functional expression of the introduced gene. Again plant molecular biologists took their cue from the microbiologists and packaged into the DNA construct not only the desired transgene, with its promoter, its on-switch, but also a gene conferring resistance to a drug that could be put into the tissue culture medium of the plant cells.

Generally a gene that codes for resistance to an antibiotic is used, a gene that causes the breakdown of the antibiotic and renders it unable to kill the cells. This allows the very few plant cells that have the antibiotic resistance gene and the linked gene of interest, in this case the Cry protein gene, to survive, divide and increase in number while the cells that have failed to integrate the insert will die.

This process of positive selection, selecting with the antibiotic for the successfully engineered cells is standard in genetic engineering. The ability to grow the plant cells and then have each surviving cell give rise to an adult plant, a technique developed many years before recombinant DNA techniques were developed, is key to the success of plant genetic engineering. Several companies have engineered corn to express a Bt toxin effective against the European corn borer using the techniques described.

They chose the toxin protein active against the ECB, generally the protein of the type CrylA(a) or (b) and used a variety of gene regulatory elements to provide expression of the protein. Also engineered into the corn were selection markers to allow positive selection of the successfully transformed plant cells after the introduction of the DNA.

The parental transgenic line is used as a parent line for the development of a hybrid for sale as seed. Before the hybrid seed can be sold and used, it must receive approval from the USDA's Animal and Plant Health Inspection Service (APHIS), from the EPA Office of Pesticide Protection's Biopesticides and Pollution Prevention Division (BPPD) and from the U.S. Food and Drug Administration (FDA) Centre for Food Safety and Applied Nutrition. Issues addressed by the USDA are the safety of the pesticidal material to nontarget organisms, including humans as well as risks to the environment. Environmental assessment and impact statements are developed. The USDA concluded the following for each Bt transgenic corn:

- The introduced genes, their products and the added regulatory sequences do not present a plant pest risk
- Expression of the insect control protein will not cause these plants to become any more "weedy" than other corn
- Movement of the introduced genes from the transgenic line into wild or cultivated sexually compatible plants is unlikely

- Use of the Bt corn should have no more adverse effects on raw or processed agricultural commodities than the parent corn
- The Bt corn exhibits no significant potential to harm organisms beneficial to agriculture, to harm threatened or endangered species or to have an adverse impact on the ability to control nontarget insect pests
- Cultivation of the Bt corn should not reduce the ability to control insects in corn or other crops

The last issue is central to the controversy about the use of Bt corn and the bureaucracy put in place to address that issue centred on the EPA The EPA worked with the petitioners to develop strategies to reduce the risk of the development of ECB and other insects resistant to Bt toxin. Ironically the concern about resistance derives from the key advantage of the Bt transgenic corn: It is designed to constantly express the Bt toxin.

This is the very method that biologists use to select for, not generate, drug resistance mutants. Mutations, alterations in the genetic material, occur with a low but finite frequency in all organisms. If environmental conditions give the mutant a selective advantage in growth and reproduction, then the mutant will replace the wild type.

This is, of course, the basis for evolution, random mutation and natural selection. With the Bt corn environmentalists were concerned that the Bt gene expressed constantly at a high level would create conditions for a very unnatural selection, and the ECB sensitive to Bt toxin would be replaced by ECB resistant to the effects of the Bt toxin.

It fell to the EPA to address concerns raised by its scientific advisors as well as public interest groups about the environmental danger of pesticide transgenic plants. In 1995 the EPA determined that plants engineered to express a pesticide would be regulated by the EPA as a pesticide. To deal with the resistance management issue they formed a Pesticide Resistance Management Working Group, which has identified the elements that need to be addressed to develop an adequate resistance management plant and has reviewed, among others, the plan voluntarily submitted by Ciba/ Mycogen for resistance management for BtCrylA(b) corn.

The issue of damage to plants, of plant pathogenicity, was raised and addressed by the USDA because genetic control elements used in the development were derived from know plant pathogens, for example, cauliflower mosaic virus (CaMV) and *Agrobacterium tumefaciens* (causing crown gall disease), but the USDA concluded that the genes used from those organisms posed no threat.

The FDA Centre for Food Safety (CFS) developed a policy in 1992 that calls for consultation between the FDA and the firm developing the new food plant varieties derived through recombinant DNA technology. The first consultation was with Calgene, at the company's request, regarding the Flavr Savr Tomato, engineered to delay softening. After evaluation of the safety and nutritional data collected by Calgene about the Flavr Savr Tomato, the FDA decided that the Flavr Savr Tomato is as safe as other commercial varieties of tomatoes.

The FDA also concluded that it was not necessary to conduct comprehensive scientific reviews of foods derived from bioengineered plants. The FDA does request that developers consult with the FDA on safety and regulatory questions. Between 1994 and 1997, thirty-one such consultations have been completed, including six for corn protected from insects with a Cry IA gene. The question is whether the food, the corn, is different from any other corn in any detectible way, because it was grown from a seed with a transgene. The FDA appears to have concluded that the means of the development of the seed is not relevant for its concerns about food safety. The product, not the process, is the issue.

The regulatory process developed by the U.S. agencies—the USDA, the EPA and the FDA—provides the necessary legal approval so that the seed developers and their sales companies can sell their Bt corn seed, and the farmers and wholesalers can sell their harvested Bt corn. However, will the corn grown from that seed actually be sold?

Advocacy groups on both sides have made known their stances. As noted earlier, the majority of corn is used in foods and beverages. If consumers are uncertain of the safety of the food, whether it is popcorn or a food (spaghetti sauce) or beverage (soda) sweetened with corn syrup, then the large food manufacturing companies will be concerned that their products will not sell, and they in turn will not buy corn from Bt seed. Corn is sold through commodity markets, organized traders' exchanges, in which contracts for delivery of certain products are bought and sold.

The Chicago Board of Trade, the world's largest futures market, publicly stated that it would accept the genetically modified corn to satisfy its contracts. The U.S. Corn Refiners Association announced that food and feed products derived from Bt corn meet all current safety regulations of the FDA, USDA and EPA in the United States and the comparable regulations of the European Union (EU).

The Institute of Food Technologists, a nonprofit scientific society with 28,000 members working in food science in industry, government and academia, stated that there is no scientific evidence of environmental or health risks associated with gene spiced corn. Others disagree. Opposition to the use of Bt corn has ranged in intensity and perspective. The Sierra Club, a well-known U.S. environmental interest group, urges caution and significant government regulation to prevent environmental damage by the introduction of genetically engineered organisms into the environment.

The Union of Concerned Scientists (UCS) raises the concern that wide use of Bt crops will lead to the development, over the course of perhaps as few as three to five years, of resistance to Bt. If resistance develops, the Bt toxin will be useless as a pesticide. Loss of Bt will affect not only those who currently use the engineered Bt crops, but also many other farmers who use Bt bacteria as a spray.

Farmers who grow food organically and those who use Bt as part of integrated pest management (IPM), will lose a useful and safe insecticide. The UCS suggests that companies that sell the Bt crops do not have the right to use up this resource guided only by commercial calculations. UCS considers Bt a public good and is working to save Bt for everyone (UCS Web site, Rissler and Mellon). UCS claims to have been instrumental in the decision by the EPA to regulate transgenic Bt and other transgenic pesticides as pesticides. This decision added another layer of regulatory process to the commercialization of Bt transgenic plants and has been criticized by many, particularly small businesses for whom the expense of money and time seeking this approval may be prohibitive.

Greenpeace, an international environmental group based in Amsterdam, has spearheaded efforts to limit the use of the Bt corn and other genetically engineered food by urging a rigorous labeling policy for genetically engineered foodstuffs in the European Union and its member countries and by initiating legal action against the EPA. The effort to require labeling is clear. "Greenpeace remains totally opposed to the introduction of genetically engineered organisms into the environment and food supplies," said campaigner Michelle Sheather. "But at the very least customers must have the right to know which food has been genetically manipulated."

In the U.S., Greenpeace has led a coalition of petitioners including the International Federation of Organic Agriculture Movements (IFOAM), the Sierra Club, the Centre for International Technology Assessment in Washington, D.C., the Institute for Agriculture and Trade Policy in Minneapolis, the National Campaign against Misuse of Pesticides and other individuals, charging the EPA with gross negligence over its approval of genetically engineered plants. The petition demands that the EPA withdraw the approval of transgenic plants carrying the genetic code from a soil bacterium called *Bacillus thuringiensis* and abstain from any new registration of such plants.

This concerted legal approach claiming regulatory negligence is reminiscent of the approach used in 1982 by Jeremy Rifkin to prevent the field testing of genetically engineered bacteria with the ability to reduce frost damage to plants. In fact, the Foundation on Economic Trends (FET), a nonprofit group led by Rifkin and opposed to much of biotechnology, has warned manufacturers that the FET will call for a boycott of foods from manufacturers who do not pledge not to use genetically modified corn or soybeans, the other major food commodity for which a genetically altered plant is currently approved.

The European Commission, the agency of the European Union charged with formulating rules for the marketing of genetically modified organisms, continues to debate how to regulate the entry of GE corn and soybeans. Trade organizations and individual countries within the EU are opposed to the entry of the GE corn and soybeans, want segregation and labeling of the GE commodities and support the imposition of an authorization process, which will significantly increase the costs for U.S. exporters. The direct impact of the decisions of the EU may not be great since, in 1996, export of corn to the EU accounted for less than 2 percent of production. But the negative publicity and the alarms raised by the opposition groups in Europe, including Greenpeace, add to the potential for alarm in the United States.

In late May 1999 environmental concerns about Bt corn were further heightened when an entomologist from Cornell University published a short paper in *Nature* describing the results of an experiment in which he fed Monarch (butterfly) caterpillars on milkweed leaves dusted with pollen from Bt transgenic corn, pollen from nontransgenic corn or no pollen. At 4 days, less than 60 percent of the caterpillars fed on milkweed dusted with Bt pollen survived, and they weighed less than half of the control caterpillars. Advocates on both sides of the debate argued about the relevance and rigor of the study, particularly whether in the field corn pollen would reach the surrounding milkweed in the amount used in the experiment and whether the caterpillars would be feeding at the time of pollen production by corn plants.

These questions have not been resolved, but this paper does provide a "poster child" for the environmentalists and adds to the negative image of the large agriculture biotech companies, such as Novartis and Monsanto. For Mr. Leary's decision, a key issue is whether Bt corn will allow him to make more money. To answer that question, three things must be known: does Bt work, will it increase his yield when compared to non-Bt corn and is that increase cost effective, considering that Bt seed is more expensive and that he may be able to avoid the expense of pesticide applications? The answers to these questions will only be certain after farmers in different regions have gained some experience with Bt seed.

Concerning the issue of whether Bt corn works, there are several reported studies comparing Bt hybrid corn seed to non-Bt hybrid of the same genetic makeup for evidence of infestation with ECB. In one study in Illinois five different Bt hybrids were compared with non-Bt corn hybrids in fields that were either manually infested with ECB or left alone for natural infestation. In all cases, fewer plants from Bt corn seed showed ECB injury and fewer larvae per plant, though the Monsanto YieldGard technology protected plants longer, particularly after pollination.

However, when yield was determined and compared, there was no significant difference in yield, expressed as bushels of corn harvested/acre, between each Bt hybrid and its non-Bt counterpart. Injury was significantly reduced for Bt corn, but yield was not increased. Not a clear recommendation.

Farmers interviewed after the 1996 harvest had mixed reviews. Farmers who had and have little corn borer damage consider the $20 to $30 increased cost for each bag of Bt corn seed as poor economics. Others who have higher risk of corn borer damage and greater damage and are likely to apply insecticide consider that the increased cost of seed, which works out to $8 to $10 per acre, is less expensive than spraying, which can cost $20 to $30 an acre.

It will take time and experience to see if farmers adopt the new seed. In the 1996 growing season, when the first Bt corn was available, Bt corn accounted for less than 1 percent of seed planted in the United States. In 1997 this figure rose to

9 percent, and by 1998 22 percent of U.S. corn acreage was planted with Bt corn. The issue of the potential for Bt corn to cause the development, or more precisely the selection, for ECB resistance to Bt has been raised from the very beginning—including public awareness of the intent to seek approval to market Bt corn and other Bt toxin transgenic plants such as cotton. Scientifically speaking there is a good basis for the concern based on theory, small-scale experiments and field experience with other insects and other Bt transgenes.

The risk, upon widespread adoption of Bt corn, to entire corn growing regions led the developers of the corn, upon realization of this risk, to develop management strategies for farmers to adopt to avoid the development of resistance to the Bt toxin used by the ECB. This is both sound environmentalism and good business practice since Bt corn would be not worth the premium price if it were ineffective against the ECB. Laboratory experiments have also shown that an insect resistant to one Cry protein may be resistant to others, so the developers could not necessarily overcome problems by using another Cry gene. Even if this general resistance were not seen, the cost of developing new transgenic strains to keep ahead of resistance, if that were possible, would be very high.

It is not possible to predict the mechanisms of resistance that could evolve in the ECB. They could be genetically dominant, that is, one copy of the resistance gene would be sufficient. The resistance gene may be recessive—two copies are needed to confer resistance. Strategies for resistance avoidance must be independent of any supposition of biochemical or genetic mechanism.

Entomologists (experts on insects) and population geneticists have considered several approaches to slowing down the development of resistance to pesticides in general and have concluded that the approach most likely to succeed in slowing the development of resistance to Bt is to plant sensitive versions of the same crop nearby to provide a refuge for Bt sensitive insects. In the absence of the Bt toxin neither the Bt resistant nor sensitive insects have advantage, and both will survive and breed. Thus sensitive plants provide a reservoir of sensitivity genes, whatever those may be. There are at least three problems with this strategy:

1. High levels of expression of the toxin is required to avoid the development of intermediate levels of resistance,
2. Refuge plants must be planted close to Bt plants,
3. The resistance genes cannot be dominant. Dominant resistance has not been seen yet. Expression of the Bt toxin is not high in all Bt corn seed hybrids and could be sufficiently active for a particular infestation level, but undermine the resistance management strategy.

The EPA, in response to alarms raised by a number of public interest groups about the issue of pesticide resistance, in 1992 created the Pesticide Resistance Management Workgroup (PRMW). The PRMW has made recommendations about what should be considered in a resistance management strategy and urged that resistance management be considered an element of EPA's assessment of the environmental impact of registration of the transgenic Bt toxin as a pesticide. Each company filing for registration of Bt corn or other plants has voluntarily submitted a resistance management programme to the EPA, and these plans have been reviewed as part of the dossier.

These plans require that farmers attend educational sessions and work closely with the company in the application of the resistance management strategy. The companies involved in developing Bt plants have also formed a Bt Resistance Management Workgroup through which they are funding research at academic institutions. The monitoring and reporting of resistance are also part of the management plans. Yet, there is really no adequate protection for the avoidance of resistance, no teeth to the EPA regulations.

The issue of resistance may seem less alarming if one considers the small amount of use of Bt pesticide; its loss to corn farmers may not be seen as catastrophic. It would mean that one nonchemical biopesticide option for corn farmers for

ECB control would be lost. The loss of certain markets, for example, the EU, may be more significant economically to the corn and corn products industry, which would have an impact trickling down to the farmer. Again, only time and experience, will tell. The issue has become more and more contentious, in the European Union and the United Kingdom, and field-testing, much less commercialization of Bt corn seed, remains controversial.

The development of Bt corn and other Bt transgenic plants can be considered an example of what Krimsky has called push (versus pull) innovation. Industry innovators developed Bt corn because they could—because it was an entry into a new technology—and they estimated that seed corn clients would not only buy it but pay the premium. ECB damage and subsequent loss in yield is not a uniformly significant problem for corn farmers, and it is not easily predictable from year to year for any particular farmer.

The need for the Bt transgene solution to the ECB problem is not absolute. Agricultural biotechnology is full of examples of push rather than pulls innovations, perhaps because the market is diverse and profit margins small. This makes the business of ag biotech as risky as the farming industry to which it sells its products. The combination of risk, expense and narrow margins also contributes to a very aggressive stance by the companies in protecting their intellectual property rights.

The companies engaged in developing Bt transgenic plants are also engaged in a complex war over patents and infringement of patent rights. Novartis (formerly Sandoz and Ciba) is suing Monsanto, DeKalb and Pioneer over the method of protecting corn against insects, including Bt protein ingestion by corn borers. Mycogen has filed suit against Monsanto and DeKalb, claiming infringement of patent claims on modification of Bt genes for plant expression, introduction of genes into plants and plants and seeds from cells transformed with Bt genes.

Plant Genetic Sciences is suing Mycogen and Ciba over infringement of rights to Bt expressed in plants. The list goes on. These suits are expensive and will add to the price of the product no matter who wins. The larger companies may use the costs and the burden and delay caused by legal wrangles to discourage smaller companies with smaller pockets and less stomach for a fight.

This may be contributing to the growth of mergers in this industry. Large multinational companies are struggling to lay claim to the tools and thus the profits of the new technology in the agriculture industry (at least to the agriculture industry in the developed world). Legal battles over ownership are just one aspect of the contest. It will take many years not only to see who wins in the courts, but also who wins in the farmers' fields, at least in the fields of the rich countries.

These expensive marginal improvements in technology do not address the needs of farmers in less developed countries, where seed at $20 to $30 per bag is unthinkable. But just as large pharmaceutical firms focus their investments in the diseases of those who can afford the treatment, the agricultural biotech industry at the level of the large multinational companies focus their investments on those segments of the agriculture production industry that can afford value added seeds.

Back in Iowa, should Mr. Leary buy and plant Bt corn seed? Not a simple question, but it will in the short run hinge on the details of the economics and his willingness to undertake self-protective resistance management efforts. He must guess at his risk of ECB for the upcoming season and weigh the cost of the seed and the cost in material and labour of applying insecticide. These calculations require a clear and cool business head.

The difference between a financially successful year and failure is in these careful calculations and guesses. In the long run he has to consider the risk of resistance development despite his efforts, the subsequent loss of the utility of Bt spray and the impact of consumer concerns about genetically engineered foodstuff. If he cannot sell his corn or the price is reduced because of market resistance, the calculation becomes more complex. Not a business for the faint of heart, this farming thing.

Chapter 6

The Development of Biotechnology

THE BACKGROUND

This chapter discusses the development of the biotechnology industry in an Italian region, Lombardy. More specifically, it asks why significant innovative and commercial activities in biotechnology did not emerge in what might have been considered at the outset a promising area for the growth of this industry, and why in very recent years some timid symptoms of dynamism seem to be appearing.

As a case study about a failure, rather than a success story, it is difficult to draw firm conclusions. In a sense, the problem is over determined. There are too many possible candidate variables that might explain why innovative activities in biotechnology did not take off and the available data do not allow any serious statistical test.

As a consequence, the chapter is largely speculative. However, we shall try to make progress in turning speculation into a substantial account, based on a theory-driven interpretation of the determinants of the patterns of innovative activities in specific geographical areas.

There is now considerable empirical evidence that innovative activities tend indeed to cluster in specific geographic areas, at different levels of aggregation. The case of biotechnology is extreme and puzzling, in this respect, from a variety of perspectives. It is very well known that the biotechnology industry is largely an American phenomenon and that other countries have been lagging behind for over two decades.

Despite some recent encouraging signals of dynamism in some European countries and in Japan, the American leadership is still undisputed. Even within the USA, the development of biotechnology has been characterised by high degrees of regional concentration. In Europe, too, innovative activities in biotechnology tend to agglomerate into specific areas. So, what are the factors that explain different performances in 'national biotech'? And what are the factors leading to agglomeration in the case of biotechnology?

After an overview of the patterns of the development of biotechnology in Italy (second section), the specific case of the biggest biotechnology cluster in Italy - Lombardy - will be described (third section). Then, we shall briefly touch upon two main conceptual issues which have attracted the attention of many analysts in recent years and which drive this chapter too, namely:

1. What are the factors that might explain the lagging behind of the Italian (and more generally, European) biotechnology industry vis-à-vis the United States (fourth section)?
2. Do innovative activities tend to cluster in specific geographical areas? If the answer is positive, what are the determinants of these processes of agglomeration, at least in the case of Lombardy (fifth section)?

In Italy, as in other European countries, innovative activities in biotechnology lagged significantly behind the USA and proceeded along different lines. Structural weaknesses in the industrial base, in the research system and at the institutional level hindered the development of biotechnology.

The main difference, of course, is the virtual absence of the phenomenon of specialised biotechnology start-ups, even as compared to other European countries. No reliable figure of the number of new biotechnology firms in Italy is available, but a rough estimate would suggest that they are around less than two dozen. Moreover, the few Italian start-ups do not resemble the American prototype in many respects. Most of them are not involved in drug research or development but are instead either intermediaries commercialising products developed elsewhere, or in the provision of instrumentation and/or reagents.

Finally, some of these companies (especially the most significant ones) have been founded as spin-offs of larger pharmaceutical companies undergoing processes of restructuring rather than as university spin-offs. However, the major distinctive feature of the Italian case is the delay with which innovative activities got started and the limited size and scope of such efforts, even within the large chemical and pharmaceutical (and agro-food) companies. Even today, Italy has to be considered a backward country in biotechnology. Just to give a rough idea, it is worth noting that the number of Italian patent applications to the European Patent Office totals only 163 in the whole period from 1978 to 1996.

In the absence of extensive new firm creation, most of the innovation in biotechnology in Italy - as in other countries of mainland Europe - has occurred within established firms and public research centres. The main actors in the development of biotechnology in Italy were the large chemical firms. The beginning of research activities, as well as the emergence of public policy, was the background to the formation of a 'biotechnology lobby', who stimulated the interest of industry and above all of the government. By the early 1980s, the two largest Italian chemical groups - ENI (Ente Nazionale Idrocarburi) and Montedison - had already started some small research activities and had considered the possibility of some more systematic effort.

However, such plans resulted only in the establishment of two small research agreements with Italian academic laboratories. Besides ENI and Montedison, only a handful of medium-sized pharmaceutical companies engaged in innovative activities, whilst the absence of any research tradition in the food industry and in agriculture prevented any significant effort in those directions. Around the chemical industry, however, a biotechnology lobby gathered and organised. Scientists were the first to perceive the potentialities of biotechnology and took an active role in eliciting both industrial and public interest in the new emerging techniques.

Yet they had traditionally few linkages with industry and their pressure achieved results only when they succeeded in involving the Association of the Chemical Industry (Federchimica). Federchimica was perhaps the single most influential institution in the development of Italian biotechnology. In 1986, it published the first major report on biotechnology and from then on it lobbied for getting public support. Subsequently, Federchimica spun off other initiatives, particularly the Italian Biotechnology Industrial Association, although its main result was to elicit and contribute to the design of government intervention.

The weakness of government support is a further distinguishing feature of the Italian case. Before 1986, public support for biotechnology was very small and unco-ordinated. It flowed mainly to universities through the National Research Council's (CNR) 'Finalised Projects', but research funding was small and it was spread over many diverse and very general projects, without any clearly defined priority. Industry was only marginally involved in these projects. Industrial research was funded through the two existing schemes supporting applied research, but again no specific priority was attributed to biotechnology.

As in most countries, the Italian government agencies knew very little about biotechnology and, given the uncertainty surrounding its future developments, they were not able to devise any specific policy. In contrast to other experiences, though, the Italian administrative apparatus also lacked the experience, the competencies and the structures for dealing with the problems of formulation and implementation of a set of policies in high-tech industries. Thus, the existence of biotechnology was not even perceived until the mid 1980s, under the pressure of the biotechnology lobby and the example set by other countries.

Attempts to co-ordinate interventions and to devise a coherent strategy started only in the second half of the 1980s, when a National Committee for Biotechnology was established at the Ministry for Scientific and Technological Research (MURST). The Committee comprised the scientists and the industrialists of the original biotechnology lobby and eventually designed the National Plan for Biotechnology, which was passed by the Parliament in 1987.

The plan was aimed at supporting applied research in different fields (especially biomedical research), while support for basic research and industry-university collaboration was provided through new programmes launched by the CNR. Other initiatives were taken to stimulate research conducted in the public research centres like ENEA (Ente Nazionale per le Energie Alternative) and the CNR. In the following years other programmes were launched by the CNR, by MURST and by other Ministries.

It is very difficult to evaluate the effective size of the financial resources devoted to the support of (basic and applied) research and to the promotion of industry-university collaboration, let alone the impact of these projects. An extremely rough estimate would suggest that in the period from 1987 to 1996 the total funding amounted to a figure around L1,000-1,200 billion.

An analysis of the National Research Plans on biotechnology launched by the Ministry of University and Scientific and Technological Research in the period from 1989 to 1997 suggests, however, that only a small group of universities and an even smaller group of companies were involved in these projects, with a very limited scientific output (only 1 patent extended outside Italy) and little interaction between universities and the companies.

In addition to these initiatives, the development of biotechnology in Italy was characterised by the mushrooming of local collaborative initiatives between public and private agents at the local level, through the establishment of science parks, research consortia, venture capital funds, etc. However, the scope and the impact of these efforts remained on the whole quite limited.

Thus, despite these efforts, very few firms engaged in significant research activities and their innovativeness was far lower than that of the major European competitors. Very few firms account for a large proportion of biotechnology patents, and innovation in biotechnology rests essentially on the activities of a small group of large/medium-sized established companies. As a consequence, Italian performance as a whole remains far behind that of Germany, France, UK, Switzerland, and also of Sweden and Denmark.

Still in the 1980s, the main actors were the two large Italian chemical groups, ENI and Montedison. As far as Montedison is concerned, research capabilities in biotechnology were present in the laboratories of its pharmaceutical subsidiary (Farmitalia-Carlo Erba, at that time the largest and most prestigious Italian pharmaceutical company) and in the central laboratories of the group (Istituto di Ricerche Donegani). In the late 1980s, Montedison expanded its biotechnology research through acquisitions and through the development of a network of collaborations with domestic universities and American laboratories and specialised biotechnology firms.

As for ENI, a significant biotechnology group was created within EniRicerche (the research laboratories of the group). Moreover, ENI owned Sclavo, a medium-sized company located in Siena, considered to be among the world leaders in vaccine production.

Finally, a subsidiary of the Fiat group, Sorin, active in the production of medical devices, started research on the development of diagnostic kits based on the use of monoclonal antibodies and assays.

Since the beginning of the 1990s, however, even these initiatives have been drastically downsized. The Italian chemical and pharmaceutical industries underwent a deep crisis. The Italian chemical and pharmaceutical companies were already structurally weak, small by international standards and their commitment to innovative activities was certainly not comparable to that of their international competitors. The major Italian chemical and pharmaceutical companies were acquired by foreign corporations (more than 70 per cent of total pharmaceutical sales on the Italian market are now generated by foreign-owned firms and by subsidiaries of large multinationals). Moreover, a series of scandals linked to the spread of a bribery system in the processes of setting drug prices, severely hit the whole industry.

Finally, cost containment policies by the Government implied a further shock to the profits and the strategies of the pharmaceutical companies operating in Italy. As a consequence, investments in biotechnology R&D were drastically reduced. For example, the pharmaceutical division of Montedison - Farmitalia Carlo Erba - was eventually acquired by Pharmacia and its research activities were subject to substantial restructuring.

Similarly Sclavo (now called Biocine), after a series of break-ups and vicissitudes, was acquired by Chiron. Many research projects were discontinued and - given the overall weakness of Italian research capabilities - only few initiatives were launched or maintained by the foreign research headquarters to compensate for the decrease in domestic investments.

The history of the development of biotechnology in Lombardy coincides to a very large extent with the history of Italian biotechnology as a whole. Indeed, since its inception, innovative activities in this area have been strongly concentrated in a few regions, Lombardy being by far the single most important centre, followed by Tuscany and - in the 1990s - by Latium. Lombardy accounts for almost half (43.9 per cent) of the patents applied for to the European Patent Office in the period from 1979 to 1996 and its contribution has been only slightly declining in the period from 1990 to 1996 (falling to 39.4 per cent). All in all, Lombardy, Latium and Tuscany contribute around 80 per cent of total Italian patenting activities and over the past few years the overall degree of spatial concentration of innovative activities in biotechnology has changed very little.

The agglomeration of innovative activities in Lombardy - and more specifically in the area around Milan - might be simply explained by the fact that Milan has the strongest concentration of research laboratories, both academic and industrial, and it is by far the most important financial centre of Italy.

Milan is traditionally one of the major centres of academic research in medicine and in biology. Whilst other universities had perhaps the best research teams and the more advanced 'schools', yet in Milan could be found a very strong concentration of units and researchers in different disciplines.

Lombardy has traditionally hosted a large majority of the chemical and pharmaceutical companies, especially their headquarters and their research laboratories. More generally, Lombardy consistently shows a very strong technological specialisation in chemicals and pharmaceuticals, whichever indicator is used (R&D expenditures, patents, number of researchers, etc.).

At the beginning of the 1990s, Lombardy hosted 49 per cent of the Italian firms active in biotechnology, 36.3 per cent of the researchers, and 42.6 per cent of the research laboratories. Similarly, Lombardy attracted 29 per cent of the public funds made available by the National Research Council's 'Finalised Project' on Biotechnology. It is also worth noting that, respectively, more than three-quarters and around 60 per cent of all the collaborations among firms - both at the national and international level - and universities and companies involved in biotechnology are located in Lombardy.

Against this background, a series of initiatives was taken to further develop biotechnology research and commercial activities, primarily through the creation of science parks or - more generally - aggregations of academic and industrial researchers to promote collaborative research and act as incubators for academic spin-offs.

The first initiative was the so-called 'Biopolo', a consortium between a few pharmaceutical companies and academics of the Faculties of Biology, Chemistry and Agricultural Sciences, with the participation - under various headings - of the local authorities. The Biopolo was an offspring of the original 'biotechnology lobby' and was promoted and stimulated by a group of professors of the State University of Milan since the late 1980s. Its main area of interest was the applications of biotechnology to the agro-food industry and environment.

In practice, the Biopolo took off very slowly, as a consequence of the scandals in the pharmaceutical industry, political meddling and academic politics. Its main activity consisted in supporting the creation of a specific field denominated 'biotechnology' within the Faculty of Sciences of the State Milan University and collaboration between university and industry was essentially focused on teaching activities and training. Only recently, with the creation of the second state University of Milan in 1998 - resulting from the splitting of the previous institution - the Biopolo has started operating in a more systematic way. With the creation of the second University, in fact, a Faculty of Biotechnology was set up which is beginning to act as a centre of aggregation for research.

A second initiative, the S. Raffaele Science Park, was more successful. The Science Park was launched in 1993 under the impetus of the head of the S. Raffaele hospital, a private institution created and managed by a charismatic (and highly controversial) priest/entrepreneur, Don Luigi Verze. The San Raffaele Hospital was conceived and developed to put a high priority on research - particularly on biotechnology and in general on molecular biology - and it quickly became a leading scientific centre in the biomedical area in Italy, with solid relationships with the international community and high-quality standards and reputation.

Around the S. Raffaele Hospital, DIBIT (Department of Biological and Technological Research) and subsequently the

S. Raffaele Science Park were formed, initially without any public funding. Now, the science park hosts sixty-three research units, including a few laboratories of the National Research Council, of the Polytechnic of Milan and, since last year, of the new University 'Vita e Salute' (another initiative of Don Verze, with Faculties of Medicine and Psychology, and conceived as a highly prestigious and selective institution). A number of important international pharmaceutical companies and a couple of new start-ups (which, however, are not yet completely operative) participate in the S. Raffaele Science Park, with specialised laboratories in a large variety of research areas. Moreover, the S. Raffaele Science Park hosts a number of laboratories funded by initiatives like Telethon and by other private foundations.

Although the S. Raffaele Science Park has among its main objectives the promotion of and assistance to spin-offs, with the provision of various types of services (scientific, technological, logistical, financial and administrative services), no really important result has been achieved so far in this area.

In fact, the most important and notable examples of new biotechnology companies in Lombardy are not university spin-offs, but industrial spin-offs, born out of the processes of restructuring some large pharmaceutical companies.

One of these cases was Lepetit, leader in antibiotics research and production. After being acquired by Marion Merrel Dow, its laboratories had become one the major research centres in this field. After a series of vicissitudes, however, Lepetit was finally acquired by HMR (Hoechst-Marion-Roussel) and the lab was closed. At this stage, a group of researchers took the initiative to launch a new company, Biosearch Italia, trying to maintain and exploit the accumulated know-how.

With the support of venture capitalists (3i-Investor in Industry), the new company was founded with the aim of being quoted in the newly created 'Nuovo Mercato' (the equivalent of the Neuer Markt or the Nouveau Marché) in the year 2000. Its main research lines are focused on antibiotics, but new areas have also been opened in anti-tumorals and genomics, in collaboration with the former parent company, now Aventis.

A similar story is that of Newron. Here, once again, a group of researchers from the Milan-based R&D laboratories of Pharmacia and Upjohn (formerly Farmitalia - Carlo Erba), closed after the change of property, decided to continue their research with funding from 3i in the area of the central nervous system. So far, Newron has been renting some spaces and services from Biosearch Italy, but is expected to be quoted on the Nuovo Mercato in 3 years' time.

NovusPharma is a further example, being born from the restructuring processes of the Boehringer Mannheim branch operating in the area of oncology. After the acquisition from Hoffman La Roche, the laboratories were closed. In this case, three European venture capital firms (the Italian branch of 3i, Atlas Ventures (NL) e Sofinnova (F)) supported the initiative of some researchers to create their own company. NovusPharma has already developed collaborations with the National Institute on Tumours, the European Institute of Oncology and the Faculty of Biotechnology of the second University of Milan and maintains an agreement with Roche. NovusPharma has ambitious growth targets and quotation on the Stock Exchange is expected in 2 years' time.

Factors Explaining the (slow) Development of Biotechnology

To a large extent, explanations of the feeble development of biotechnology in Italy, of the clustering of innovative activities in Lombardy (and a few other areas), and of the specific evolution of the industry in Lombardy tend to coincide. We would also advance the hypothesis that the Italian case is just an extreme instantiation of a more general European experience.

To deal with these questions, however, two different issues must be distinguished. First, why did NBFs not flourish in Europe? Second, why were European large established companies - in general terms - slower than their American counterparts in adopting biotechnology - more precisely molecular biology?

Factors hindering the development of new biotechnology companies

It is now widely recognised that in the United States a combination of factors made it possible for small, newly founded firms to take advantage of the opportunity created by biotechnology. In Europe (although to a lesser extent in the UK) many of these factors were not in place. Let us briefly examine the role of these factors in the case of Europe, and Italy in particular.

Strength of the Local Scientific Base

It is widely recognised that the very strong state of American academic molecular biology played an important role in both facilitating the proliferation of start-ups. The strength of the local science base may also be responsible, within Europe, for the relative British advantage and the delay of France and Germany. There seems to be no doubt about the superiority of the American and British scientific systems in the field of molecular biology, and it is tempting to suggest that the strength of the local science base provides an easy explanation for regional differences in the speed with which biotechnology developed.

In Italy, it is widely acknowledged that the level and the scope of basic research in molecular biology was not very high. Indeed, some research groups conducting state-of-the-art research were present in the early 1980s, especially in fields like immunology. Naples and Pavia were particularly important in this respect. In Naples, IIGB (International Institute of Genetics and Biophysics), which was directed for a few years during the 1960s by James Watson, was an internationally renowned research centre that spun off successive generations of high-quality researchers, and clusters of research groups spread all over Italy.

Similarly, Pavia constituted another pole of excellence, where Italian studies in genetics were practically born thanks to the work of Buzzati Traverso and where, later on, an important National Research Council group was created and developed. Milan and Rome had also some excellent research groups, whilst Genoa had a tradition of research in immunology and IST (Institute for Research on Tumours), a spin-off of the National Institute for Cancer Research, was created. Padua, finally, had developed significant competencies in the field of the chemistry of proteins.

Other research initiatives were developed within public research centres (like the CNR's groups or ENEA) and by private institutions. For example, the Italian Association for Cancer Research (mainly a funding agency) introduced a new, competitive approach to the funding of scientific research in Italy. The National Institute for Cancer Research and its offspring IST (Institute for Research on Tumors in Genoa) operated at the frontier of knowledge in experimental oncology and immunology and had a dense network of relationships with American and British laboratories.

Moreover, since the mid-1980s both institutions started a systematic attempt to slow down the brain drain process that was hitting the domestic research system, by recruiting expatriated Italian researchers. The Mario Negri Institute for Pharmacological Research, which was initially located in Bergamo and then opened up various branches in other locations, was another centre of excellence in research.

In sum, while some good research centres existed, on the whole Italian research in molecular biology and in other disciplines relevant to biotechnology was lagging behind most of the major European countries. Above all, research was concentrated in a few heterogeneous clusters, with little communication and co-ordination except personal and informal relationships among the individual researchers. Over the years, the level and scope of basic research have significantly improved, thanks also to the activities and the international contacts of a younger generation of researchers who studied and worked initially in the USA and other countries, but they still lag significantly behind the main European countries in terms of absolute amounts of funding, organisational efficiency and overall quality of the research.

The strength of the local knowledge base also goes a long way to account for the agglomeration of innovative biotechnological activities in a few clusters in Italy. However, some qualifications are needed. First, as we have seen, Milan did not necessarily have the best scientific research in absolute terms as compared to other Italian locations. Second, centres which had poles of excellence in scientific research did not become important biotechnology clusters, e.g. Naples. This would suggest, at this stage, that not only the absolute quality of science, but also the existence of critical thresholds in the 'quantity' of good science were necessary conditions for the emergence of clusters.

Industry-university Relations

An explanation based only on the strength of the local scientific base might seem unsatisfying to the degree that academic science is rapidly published and thus, in principle, rapidly available across regions and across the world. However, even in this extreme case of a science-based technology, in the early years of the industry the exploitation of 'biotechnology' required the mastery of a considerable body of tacit knowledge that could not be easily acquired from literature.

Thus, it has often been suggested that the flexibility of the American academic system, the high-mobility characteristics of the scientific labour market, the willingness to exploit the results of academic research commercially and, in general, the social, institutional and legal context that made it relatively straightforward for leading academic scientists to become deeply involved with commercial firms were also major factors in the health of the new industry.

In contrast, links between the academy and industry - particularly the ability to freely exchange personnel - appear to have been much weaker in Europe. Indeed, the efforts of several European governments were targeted precisely towards the strengthening of industry-university collaboration.

Thus, one observes a mushrooming of initiatives all across Europe aiming at establishing stronger links between industry and universities and to encourage a more entrepreneurial attitude of the universities. Policies have practically been targeted mainly at the setting up of specific organisational devices to manage technology transfer, like science and technology parks or other agencies for technology transfer. These initiatives have taken a wide variety of forms and show a mixed record in their performance.

In the case of Italy, the structure and standards governing the academic system certainly did not favour the entrepreneurial exploitation of university-based research. Interactions between industry and university had been traditionally weak and in any case largely informal and based on personal - rather than institutional - relations.

This state of affairs has changed very slowly over the years and it has been only very recently that a different attitude has started to emerge. In a survey carried out in 1994 among the academic researchers participating in the National Plan for Biotechnology - a programme launched by the Ministry of University and Scientific and Technological Research (MURST) to sustain biotechnology research and industry-university collaboration - a large fraction of the respondents to the questionnaire declared themselves willing to engage more directly in the commercial exploitation of their research.

However, administrative rules and the lack of appropriate organisational structures within the universities were quoted as the main factor hindering such developments. A more recent study on industry-university relations in Italy confirmed that, despite a renewed dynamism in this field, only a few universities have developed any systematic organisational structure to deal with relationships with industry and with the commercial exploitation of scientific research. More than 60 per cent of the universities do not even have standard contract typologies, let alone dedicated technology transfer offices.

Thus, the traditional separation between academia and industry, and the organisational deficiencies within universities make it very difficult to commercialise academic research. However, it has to be stressed that, given the weakness of scientific research in general, the absence of appropriate 'transfer mechanisms' appears to have played a relatively minor role so far. To put it provocatively, transfer mechanisms are largely irrelevant if there is nothing to be transferred. Moreover, the development of organisational structures supporting technology transfer appears to be largely dependent on the existence of a strong research base.

Access to Capital

It is commonly believed that the lack of venture capital restricted the start-up activity of biotechnology firms outside the USA. True, venture capital played an enormous role in fueling the growth of the new biotechnology-based firms (or 'NBFs'). However, at least in Europe, there have been many other sources of funds available to prospective start-ups (usually through government programmes). In addition, the results of several surveys suggest that financial constraints did not constitute the main obstacle for the founding of new biotechnology firms in Europe.

In addition, although venture capital played a critical role in the founding of US biotechnology firms, collaborations between the new firms and the larger established firms provided a potentially even more important source of capital. Yet large European companies tended to collaborate relatively more with US biotechnology firms than with prospective European start-ups. Even in the absence of other institutional barriers to entrepreneurial ventures, start-ups in Europe might have been crowded out by the large number of US-based firms anxious to trade non-US marketing rights for capital. Given the number of US NBFs in search of capital, European firms had little incentive to invest in local biotechnology firms.

As a partial support for this interpretation, we recall that in Italy several initiatives by both domestic and foreign investors to launch venture capital funds were attempted in the 1990s, with little success so far. These funds, if anything, ended up investing in new biotechnology companies outside Italy. Conversely, some of the few successful Italian NBFs have been funded by foreign venture capital firms. Thus, the failed development of venture capital in Italy seems to depend less on the lack of investors and funds than on the paucity of supply of promising start-ups based on solid scientific research.

Intellectual Property Rights

It is widely acknowledged that the establishment of clearly defined property rights also played a major role in making possible the explosion of new company formations in the USA, since the new firms, by definition, had few complementary assets that would have enabled them to appropriate returns from the new science in the absence of strong patent rights.

In the USA, a very tight appropriability regime in the biotechnology industry emerged quite quickly. In 1980, the Congress approved the so-called Bayh-Dole Act, which gave universities (and other non-profit institutions, as well as small businesses) the right to retain the property rights to inventions deriving from federally-funded research. In 1980, the US Supreme Court ruled in favour of granting patent protection to living things, and in the same year the second reformulation of the Cohen and Boyer patent for the rDNA process was approved. In the subsequent years, a number of patents were granted establishing the right for very broad claims. Finally, a one-year grace period was introduced for filing a patent after the publication of the invention.

In Italy, as in Europe as a whole, the grace period introduced in the USA is not available: any discovery that has been published is not patentable. Moreover, the interpretation has prevailed that naturally occurring entities, whether cloned or uncloned, cannot be patented. As a consequence, the scope for broad claims on patents is greatly reduced and usually process rather than product patents are granted. A draft directive from the Commission that attempted to strengthen the protection offered to biotechnology was approved by the European Parliament only in 1998, and an earlier version was rejected in 1994. Still, considerable controversy surrounds this issue.

While it is clear that stronger intellectual property protection is not unambiguously advantageous, it is, however, highly plausible that at least in the early days of the industry the USA reaped an advantage from its relatively stronger regime.

A Regulatory Climate that did not Restrict Genetic Experimentation

Opposition to genetic engineering research by the 'Green' parties is often quoted as an important factor hindering the development of biotechnology, especially in Germany and in other Northern European countries. Public opposition to biotechnology is also said to have been a factor behind the decision of some companies to establish research laboratories in the USA. In Italy, opposition from these quarters was not, however, an important factor, at least until last year, when systematic opposition to genetically modified food started to develop among consumers.

Among Large Established Firms

A further important feature of the patterns of development of European biotechnology is the delayed response of many large established companies to adopt the new techniques as compared to American (and to some extent British) companies. The relevance of this factor can hardly be underestimated. Given the low rate of creation of new start-ups, the development of biotechnology in Europe rested on the activities of large companies. The delay in the adoption of molecular biology within large corporations was therefore an essential factor explaining the European sluggishness.

Moreover, in the absence of vibrant research activity by large firms, new prospective start-ups lacked an essential source of survival and growth, through the establishment of collaborative agreements.

As mentioned previously, in the absence of such competencies, large companies would turn to the American scientific and technological base to tap and absorb the new requisite competencies during their catching-up process. Thus, in Europe a vicious circle between the relative backwardness of large firms and low rate of formation of new start-ups was created.

Firms' size, Market Structure and the Competitive Environment

The rate of adoption of molecular biology by established companies varied widely across the world and across firms. Within Europe some large British and Swiss firms were able to adopt the technology rather quickly. Other firms that had smaller research organisations, that were more local in scope, or that were more orientated towards the exploitation of well-established research trajectories, found the transition more difficult. Thus, many of the smaller American companies, and almost all of the established French, Italian, German and Japanese companies appear to have been slower to adopt the new technologies.

The relative strength of the local science base appears to be again an important part of the explanation. American and UK science was more advanced than mainland Europe: hence the slow diffusion of the new techniques to European pharmaceutical firms. This explanation is not the whole story, however. For example, many Swiss companies established strong connections with the US scientific system, suggesting that geographic proximity as such played a much less important role in the diffusion of molecular biology.

A second possible explanation is that diffusion was shaped by the relative size and structure of the various national pharmaceutical industries. The pre-existence of a strong pharmaceutical national industry, with some large internationalised companies, may have been a fundamental prerequisite for the rapid adoption of molecular biology. In many European countries (particularly in Italy), the industry was highly fragmented into relatively small companies engaged essentially in the marketing of licensed products and in the development of minor products for the domestic markets. However, size or global reach may have been a necessary condition, but the delay of the largest German companies to adopt these techniques suggests that it was not sufficient.

Thus, the most plausible explanation is that other institutional variables were also important. A possible candidate is the stringency of the regulatory environment. There is now widespread recognition that the introduction of the Kefauver-Harris Amendments had a significant impact in inducing a deep transformation of the US pharmaceutical industry, particularly through raising the cost and complexity of R&D. Partly as a result many US firms were forced to upgrade their scientific capability.

Similarly, Thomas (1994) has suggested that the European country whose leading firms did move more rapidly to adopt the new techniques - Britain - also appears to have actively encouraged a 'harsher' competitive environment. Ever since the 1960s, the British system encouraged the entry of highly skilled foreign pharmaceutical firms, and a stringent regulatory environment also facilitated a more rapid trend towards the adoption of institutional practices typical of the American and Swiss companies: in particular, product strategies based on high-priced patented molecules, strong linkages with universities and aggressive marketing strategies focused on local doctors. The resulting change in the competitive environment in the home market induced British firms to pursue strategies aiming less at the fragmentation of innovative efforts into numerous minor products, than at the concentration on few important products that could diffuse widely into the global market. By the 1970s, the ensuing transformations of British firms had led to their increasing expansion into the world markets.

The diffusion of the new technologies varied, however, not only across regions, but also across firms. Most firms that rapidly adopted the new techniques were large multinational or global companies, with a strong presence, at least as research is concerned, in the USA and generally on the international markets. Zucker and Darby (1995) present some evidence that size alone is a reasonable predictor of adoption, at least in the USA.

Henderson (1994) and Henderson and Cockburn (1996) have shown that those firms that adopted molecular biology faster and more efficiently had also already adopted the techniques of 'rational' drug discovery. By and large these were larger firms that had early developed a 'taste' for science and that were able to integrate the new knowledge within the firms. In turn, this was accomplished by a series of organisational changes directed towards building and sustaining tight links to the public research community, essentially through the successful adoption of particular, academic-like forms of organisation of research. Here other institutional factors appear to have been a necessary, albeit not a sufficient condition.

First, it is likely that the Anglo-Saxon forms of corporate governance made it easier for companies to 'hire and fire' personnel or rapidly cut non-performing assets, as compared to firms located in 'stakeholder' economies such as Germany. In addition to the problem of cutting established research competencies, European companies faced the risk of giving long-term employment to biologists before biology was proven to be successful over the long run.

Second, to the extent that the adoption of the new techniques and of the associated academic-like forms of organisation of research within companies was faster in companies that had early developed a 'taste for science', it is tempting to suggest that the origin of the American advantage in the use of biotechnology lies in the proximity and availability of first-rate scientific research in universities and in the closer integration between industry and the academic community, as compared to other countries.

One might also speculate that this was - at least to some extent - the result of the stronger scientific base of the American medical culture and - relatedly - of the adoption of tight scientific procedures in clinical trials. Through this mechanism, American companies might have to develop earlier and stronger relationships with the biomedical community and with molecular biologists in particular. Segregation of the research system from both medical practice and from close contact with commercial firms (as in France and possibly in Germany) has been highlighted as a major factor hindering the transition to molecular biology in these two countries.

Clustering

The previous observations might bear some relevance for an understanding of the causes of the formation of (and of the failure to develop) innovative clusters in high-technology industry and in biotechnology in particular.

With specific reference to the case of Italy and Lombardy, it is quite clear that most of the factors that account for the dominance of the US industry were simply not in place. Italy lacked a strong scientific base, its industrial base was weak and there was little interaction between industry and universities. Moreover, financial and regulatory constraints have plausibly hindered the development of biotechnology. Against this background, it is then not particularly surprising that innovative activities tended anyhow to cluster in those regions where at least some of these conditions were present. If anything - as mentioned above - Lombardy (and in part Tuscany in pharmaceuticals) had a stronger industrial base and it was a relatively big centre of academic research.

But are these factors sufficient or only necessary conditions for clustering to occur? And which other forces might explain the phenomenon? The question is particularly important in the case of biotechnology, which is usually considered as a strong science-based technology and as such - at least in principle - in large part abstract and codified knowledge. Under these circumstances, knowledge should be in principle available to everybody.

So, what forces led to the agglomeration of biotechnology activities in specific clusters? Part of the answer lies in the observation that mastery of biotechnology required (and still requires) a lot of complementary tacit knowledge. To the extent that the transmission of tacit knowledge is facilitated by geographic proximity, clustering may be a likely outcome. However, in the case of biotechnology other authors have suggested that the US start-ups were not simply the result of geographic proximity.

To put it in a different way, certainly geographical proximity is not a sufficient condition for the development of clusters. So, what other factors might be responsible for the observed patterns? And is clustering a necessary condition for the development of biotechnology? Recently, international trade theorists and industrial economists have rediscovered what urban and regional economists have long taken for granted about this issue. For example, Henderson lists four sources of location externalities, three of which come directly from Marshall and can be found more or less unaltered in Krugman:

1. Economies of intra-industry specialisation: a localised industry can support a greater number of specialised local suppliers of industry-specific intermediate inputs and services, thus obtaining a greater variety at a lower cost;
2. Labour market economies: a localised industry attracts and creates a pool of workers with similar skills, which benefits both the workers and their employers;
3. Ease of communication among firms: information about new technologies, goods and processes seem to flow more easily among agents located within the same area, thanks to social bonds that foster reciprocal trust and frequent face-to-face contacts. Therefore adoption, diffusion and innovation seem faster and more intense in geographical clusters than in scattered locations. That is, some 'knowledge spillovers' exist, which are geographically bounded;

4. Public intermediate inputs: local authorities may place a stronger than usual effort in providing them as soon as they recognise the importance of a specific industry for the welfare of the local communities.

All these agglomeration forces can play a role in defining a cluster of innovation. Definitely, points a), b) and to a lesser extent d) played a role in the case of Lombardy. Thus, localised innovation advantages may arise even in the absence of agglomeration force c). However, this is precisely the force that has attracted the attention of most economists and fascinated many policymakers, especially in the case of biotechnology and with specific reference to university-industry relationships.

The roie of geographically bounded knowledge spill-overs is linked to tacitness. Tacitness implies that personal contacts, imitation and frequent interactions are necessary tools for knowledge transmission, and ones which are clearly available at a lower cost for firms located within the same city or region. Regional economists have built upon such observations, and added many socio-economic features to the need for agglomeration. In particular, they have pointed out that the transmission of tacit knowledge requires mutual trust, a sharing of language and culture, as well as intense non-business relations. Thus social networks, such as those one can find in areas with a homogeneous social background, appear to be key carriers of tacit knowledge.

However, very little is known about the precise ways by which knowledge is actually transferred among people located in the same geographic area. The usual story assumes that by being near to universities and other innovative firms, employees of local firms will hear of important discoveries first and thus be able to utilise them before others are aware of their existence.

Some authors have argued that the standard notion of geographically localised knowledge spill-overs does not seem to apply to the case of biotechnology industry, at least in the phase of its emergence. Rather, they argue that discoveries in this technological area are characterised by high degrees of natural excludability, i.e. that the techniques for their replication are not widely known and anyone wishing to build on new knowledge must gain access to the research team or laboratory setting having that know-how. In these circumstances, the discovering scientists ('superstars') tend to enter into contractual arrangements with existing firms (contract or ownership) or start their own firm in order to extract the supra-normal returns from the fruits of their intellectual human capital. Moreover, the scientists work with or create a new firm within commuting distance of home or university (where they tend to retain affiliation), thus creating localised effects of university research.

These authors show that the innovative performance of biotechnology firms is positively associated to the total number of articles by local university 'star' scientists. However, when the number of articles written by university stars is broken down into those written in collaboration with firm scientists ('linked') and the remaining ('untied'), the coefficient on articles written by local university stars not in collaboration with the firm loses its significance and nearly vanishes in magnitude.

Previous evidence on the existence of indiscriminate localised knowledge spill-overs seems therefore to have resulted from a specification error, i.e. the inability to control for the actual relationships linking individual scientists to individual firms. These results suggest some conjectures:

1. At least in the early phases of new industries, knowledge is not 'in the air', but is embodied in individual scientists and research teams. Social ties and personal contacts are not sufficient to gain access to naturally excludable knowledge. This requires deep involvement in the research process and bench-level scientific collaboration. If anything, this result tends to support the idea that one must invest resources not simply to search for new knowledge, but to build the competencies to absorb the knowledge developed by others and to understand the highly context-specific 'codes' into which knowledge is translated.

2. Naturally excludable and rivalrous knowledge does not spill over, but people (teams) embodying knowledge move (locally) across organisations in order to exploit the value of their knowledge. In other terms, localised effects of university and industry research are most likely to result primarily from a combination of appropriability of tacit non-replicable knowledge and low geographical as well as organisational mobility of researchers.

   It is also worth mentioning that the assumption that inventors and firms operating in a specific industry and in close spatial proximity to each other will be more innovative than those who are geographically isolated because of the greater likelihood that they will share (tacit) knowledge has been criticised by Lamoreaux and Sokoloff (1997). In summary, they show that:

   a. Concentration of firms and production in a given area is not *per se* a necessary and sufficient condition to determine high rates of innovative activity. To put it in slightly different terms, static externalities related to the current scale or size of an industry in a given city does not necessarily generate better (local) information flows to the advantage of innovative activities. What seems to matter more is the accumulated stock of knowledge (dynamic externalities) in a diversity of industries as well as the levels and types of human capital in a region. Regions that first emerge as centres of innovative activity in a certain industry tend to maintain their advantage over time.

   b. Industries may move across regional and national borders without a corresponding re-location of inventive activity. Inventive activity is 'stickier' than production. Of course, it remains to be seen why it is so. On the one hand, there is the possibility that the locational stickiness of inventive activity derives from the reluctance or lack of incentives to migrate by people with knowledge and experience in an industry. On the other hand, a possible explanation could relate to the richness of general technological know-how in higher-order regions that serve as an effective substitute for specific knowledge and allows new applications to be found across a wide range of industries.

c. Institutions matter for regional innovation, but in a different way than frequently claimed by many regional economists. These latter tend to stress the role played by 'soft' institutions like trust, norms, codes of communication, conventions, in facilitating the process of information sharing among firms and individuals. According to another perspective, institutions are also important because they help to build those 'bridging' market mechanisms that mediate relations among inventors, suppliers of capital and those who are willing to commercially develop or exploit new technologies.

Finally, empirical evidence has been recently collected to show that linkages bounded by proximity are not always key to innovation. The results obtained show quite convincingly that local sources of knowledge are key in determining success in the development of new products and processes only in areas with a large accumulation of knowledge.

On the contrary, innovations in firms located in cities with a relatively small accumulation of knowledge depend on the relationships with universities and other high-technology firms (suppliers and customers) located elsewhere, especially in higher-order urban centres. More specifically, the results of this study permit one to draw a number of important conclusions:

1. Local boundaries play a fundamental role for the recruitment of a skilled workforce and technical personnel. The presence of other firms working on similar or related things is important not only as a direct source of engineers and scientists, but also because of the indirect effect of reducing the likelihood that a

skilled worker will suffer a long period of unemployment. In other words, local areas are important also for non-knowledge-related agglomeration economies.

2. The most dynamic and innovative firms look for knowledge embodied in engineers and scientists wherever they are available, and not necessarily constrained in this by geographical barriers. The question on the importance of intellectual human capital (engineers and scientists) developed within the metropolitan area was given very low rates by fast innovators on average, whereas they scored very high on the question of the availability of frequent flights connecting to other high-tech regions.
3. Local knowledge sources are relatively less important for firms located in lower-order regions. For these firms, local universities are viewed as suppliers of a skilled workforce, rather than loci of innovations or sources of product ideas or spill-over effects. In order to sustain high rates of innovation they must develop linkages with actors (universities and other high-tech firms) located in higher-order regions.

Trying to draw some conclusions from this discussion, it would appear that clustering may be the outcome of different factors, but mainly:

1. The existence of a strong critical mass of scientific knowledge, in absolute terms: in other words, excellence in scientific research is a basic precondition for attracting innovative activities. Without this, firms (incumbents and/or prospective entrepreneurs) might look for other locations for tapping the relevant knowledge. Moreover, diversity is also important. Insofar as innovation rests on the integration of different fragments of knowledge, the presence of a diversified scientific base becomes a key issue;
2. The existence of a strong and diversified industrial base, with accumulated capabilities and organisational

structures enabling them to actually participate in the network of cognitive and social relationships that are necessary to get access to, to absorb, integrate the new knowledge and, on these bases, to engage in successful innovative activities;

3. The existence of specific and often formal organisational devices (including markets for know-how) that allow flows of knowledge to take place.

In this chapter, we have argued that the sluggishness of the development of biotechnology in Italy was due to the lack of most of the basic preconditions for the take-off of innovative activities in this field, as it concerns the scientific and industrial base, the organisational structures linking science to industry, venture capital and intellectual property rights.

Innovative activities clustered, but did not take off, in those regions, primarily Lombardy, where at least some of these preconditions were in part present: mainly as regards the availability of good science and a solid industrial base. Such conditions were, however, not sufficiently developed. As a consequence, biotechnology did not take off.

Finally, we have also argued that clustering is a likely, but certainly not sufficient, condition for the development of innovation in biotechnology and perhaps not even a necessary condition. A cluster may develop if all the above-mentioned preconditions are in place and if 'hard' organisational devices for sustaining intense knowledge flows are created.

These observations might have some implications for policies. Indeed, in recent years European biotechnology seems to have found a new dynamism, and even in Italy and in Lombardy some interesting developments are taking place.

One interpretation of this dynamism might be that policies might - at last - have begun to exert some impact. In many countries, measures have long been taken to introduce within Europe some (or parts of) of the typical American institutional features which have been crucial to the development of new biotechnology start-ups. Thus, policies have been targeted at fostering venture capital, at developing financial markets tailored for new high-risk companies, at promoting the commercialisation of academic research and the mobility between academia and commercial activities.

This is certainly the case of Germany and perhaps of France, too. In other countries, however, the effect of policies seems to have been much smaller. And even in the German case, the effects of BioRegio have been widely different across regions. This, in turn, may be the outcome of different policy designs and/or of other structural conditions. In general, there seemed to be minimum critical thresholds that policies have to overcome before they can have some impact, and that in any case interventions aimed at modifying institutions and deeply ingrained forms of behaviour take time.

So, changes in other structural conditions may have played an important role. A primary candidate is to be looked at in a general process of maturation and catching-up in the scientific, technological and organisational base of European biotechnology, mainly as it concerns basic research in universities and in large established companies. As argued in the chapter, the strength and width of the knowledge base, both in universities and in industry, and its integration in efficient organisational forms, was and remains perhaps the single most important factor in explaining performance in biotechnology.

Thus, the increasing diffusion of molecular biology in the general training of doctors and in the medical practice, an increased integration of research and teaching, the widening and the increasing integration of the research base in Europe produces now state-of-the-art knowledge that can be exploited by NBFs, at least in specific niches, and by large corporations which are now in the position to express a qualified demand for such research and services.

To be sure, the gap with the USA is far from being filled. Moreover, the institutional structure of most European countries remains profoundly different from the American one. Thus, the future of the new European NBFs and of the patterns of evolution of the biotechnology industry remains to be seen. However, some tentative implications might be drawn from this case.

First, policies can and do have effects, especially if they overcome some critical threshold and they are 'integrated', i.e. they act simultaneously on all the relevant variables. Second, however, policies take time before exerting their effects. Third, the development of a strong research base and of strong technological competencies remains a crucial precondition for industrial dynamism and also for the efficacy of other policy initiatives aiming at inducing institutional and cultural changes: the creation of appropriate incentive mechanisms is clearly an important issue, but incentives without competencies may be totally ineffective or even dangerous.

Chapter 7

The Situation of Biotechnology

AN ASSESSMENT

The success in constructing a flourishing information and telecommunication cluster has encouraged Finland to seek equal possibilities in other knowledge-intensive industrial branches. Currently, biotechnology is seen as an auspicious business area. As in other parts of Europe, Finnish biotechnology concentrates mainly on pharmaceutical industries. More than 70 per cent of the human resources of Finnish biotechnology industries is located in that industrial area.

Despite large investments and financing, modern biotechnology industries is still a marginal industrial branch in the Finnish economy. Its turnover in 1999 was only 1 per cent of the total GNP. The wide spectrum of Finnish higher education and the close relations of co-operation between universities and industries can open up promising prospects. Finland's technology-affirmative attitudinal climate works in the same direction. This progress can be delayed by a lack of highly skilled human resources, which is already a problem in the IT sector. The small size and minor experience in global markets have also limited the success of the Finnish firms.

Finland is largely known as a country of forest industries. Until the end of the 1950s, about 90 per cent of exports from Finnish industry came from the forest or wood processing industry. In the 1960s, this figure decreased to nearly 70 per cent, but the forest cluster, constituted by forestry, mechanical forest industry, chemical forest industry, and several related supplier and customer industries, was still by far the biggest exporter in Finland. The share of the metal industry at that time was about 13 per cent.

During the following decades, the metal industry improved its position as an export industry, mainly due to the increasing trade with the Soviet Union. Nowadays the share of the forest cluster in the Finnish export industry is still 31 per cent, but the share of metal industry increased to 45 per cent. Half of the metal industry's export consists of the products of the new information and telecommunication sector. During the 1990s, Finland was able to develop the IT industry as the third pillar of its exports beside the old wood processing industry and the traditional metal industry.

Indisputably, Nokia is the dominating company in the Finnish IT sector. The share of Nokia in the recent export boom of Finnish industry became so large that fear of a 'Finnish Nokia dependence' was expressed. On the other hand, the high standard of expertise and know-how created by Nokia can be viewed as a base for constructing other high-technology clusters. Both arguments are mentioned when the need for a fourth pillar for the Finnish export industry is being discussed. In this debate, biotechnology is seen as a promising new growth area to which Finnish public attention is now very strongly drawn.

There is great hope that the success story of the IT sector, which for good reason can be considered the result of national planning can be repeated in the biotech sector. The progress made in the IT branch can be explained by a close co-operation between industry, public R&D research institutes and the education system. Business enterprises in the sector have invested heavily in R&D and at the same time about half of all the national research input is used in or for the benefit of this single branch. And in higher education, about 35 per cent of all the university and polytechnic graduates have an education in IT or neighbouring fields.

Certainly, there are other industries which are seen as possible candidates for future growth, such as the content industry and the new media industry. However, biotechnology seems to be the first choice in political as well as public debate when the advancement of a new growth area is discussed. One indicator of the fact that biotechnology is given highest priority is the increasing financial support for research in this sector.

THE FINNISH VIEW OF BIOTECHNOLOGY

In Finnish publications, the interpretation that biotechnology represents a new type of technology predominates. Up to now clustering of biotechnology takes place mainly within existing industries such as the agro-food industry, pharmaceutical industry, and wood processing and chemical pulp industry, as well as in the debilitated waste management industry. The way Kreiner and Schoulz have described the 'pre-clusterical' situation of biotechnology industries also applies to Finland:

In a loose sense, we may talk about a biotech community, since across all differences, all the competing classifications being applied and used, they identify themselves in certain situations and contexts as working in biotech.

One sign of the development of an independent biotechnology cluster, however, is the foundation of the Finnish Bioindustries (FIB) organisation. The majority of the companies in this field are now members of this association (including some foreign members), which are mostly micro enterprises; whether biotechnology will ever become an industrial cluster of its own, however, is controversial. So far, biotechnology is de facto a new method used by various industries in research, development and production, and in many different industrial clusters. In this respect, it is comparable to telecommunication technologies. The complex nature of biotechnology makes it difficult to assess its economic impacts. Nevertheless, it is possible to identify the focus domains of modern biotechnology.

In Finland, modern biotechnology is mainly applied in the pharmaceutical industry, biomedicine and bioremediation. Exploitation of modern biotechnology in the agro-food industry, and the forestry and chemical pulp industry is just taking off. Research and development work with biocatalysts and enzymes seems to produce some promising results. Many of these biocatalysts can be used in the processes of the agro-food industry and the chemical pulp industry. The waste management industry in Finland is rather underdeveloped, although some applications based on modern biotechnology are currently being researched.

Some Conceptual Aspects

For a long time, the scientific debate on the development of new technologies and new industries based upon them has been dominated by the controversy between the 'knowledge or technology-push approach' and the 'market-pull approach'. Nowadays there is a widespread agreement that both factors, newly created knowledge and the market, are important for the development of new technologies and industries. There is also, however, a third factor which must be mentioned here: the closeness of the existing knowledge stock within firms to the newly emerging knowledge.

Concerning the knowledge or technology aspect, not only the amount and the quality of research plays an important role, but the co-operation of knowledge producers, mainly universities, with industries is also decisive, since successful innovation processes depend upon an inclusive and rapid knowledge flow between these two parts of the idea-innovation chain. Furthermore, companies need to have an absorptive capacity to be able to use the new knowledge produced in the university. Their knowledge stock must be compatible with the newly produced knowledge, as learning in companies takes place in a cumulative way.

Regarding the market aspect, close co-operation between producers and users is crucially important for firms in order for them to be able to develop products for which market demand exists. In this respect, small countries with closely networked small and medium-sized firms may even have some advantage; but, in the long run, the development of new technologies that can become the basis for new industries depends upon the access to international markets through global players. In the following, we will discuss the extent to which the emerging biotechnology sector in Finland can rely on such favourable preconditions.

The Importance of Biotechnology for the Finnish Economy

So far, the importance of modern biotechnology for the Finnish economy is still limited. In 1998, the turnover of the related industries was estimated to be FIM 7,425 million (€1,249 million), which is only 1.1 per cent of the total GNP (Gross National Product). The employment rate of the Finnish modern biotechnology industry is also marginal. It employs 5,610 people, which is 1.3 per cent of the employment in the whole industry, and 0.3 per cent of the total employment. About 4,000 people employed in the sector are working in the pharmaceutical industry, which is more than 70 per cent of the workforce in this sector.

Biotechnology-related employment in process industries, including agro-food industries and enzyme production industries, amounts to 1,400 people. And about one hundred people are working in knowledge-intensive business services in the field of biotechnology. Also, the number of companies in the biotechnology sector is rather small in Finland. There are only about 120-140 companies in the sector (Finnish Bioindustries 2000), while the total number of companies in the Finnish economy is about 210,000. Most of the companies in this field are very small, employing less than ten people.

Upto now, modern biotechnology has had no significance in Finnish export statistics. In 1997, the value of exports of high-tech pharmaceuticals was only FIM 140 million (€23.5 million), and the value of exports of high-tech chemicals was FIM 310 million (€52.2 million). The combined value of exports of both high-tech industries was only 1.3 per cent of the total high-tech exports in Finland. Export of electronics and telecommunication, on the other hand, the strongest high-tech sector in Finland, had a value of FIM 22,760 million (€3,800 million).

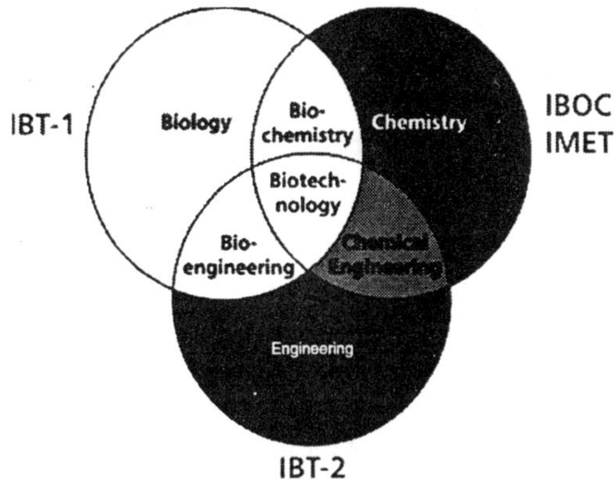

Fig. Disciplines of Biotechnology

The number of patents can also be seen as a good indicator of competitiveness. Here we can see a dramatic increase in all patent categories.

The distribution of venture capital investments represents another good indicator in addressing the importance of a particular industry. In Finland, the total value of venture capital investments in 1997 was FIM 5,850 million (€984 million), of which two-thirds were investments from the private sector. The venture capital investments in biotechnology were FIM 25 million (€4.2 million), which is only 0.4 per cent of total investments.

Despite the marginal position of modern biotechnology in the Finnish economy, there are high expectations concerning the future development of this new sector and its contribution to the Finnish economy and to Finnish exports. These expectations are supported by the results of a survey analysis conducted in Finland in 1996. According to this survey, one-fifth of the Finnish population accepts the GMO (genetically modified organism) food without reservation. Finnish people also associate some advantages with modern biotechnology.

In an EU survey, two-thirds of Finns were optimistic concerning the implications and consequences of modern biotechnology, while on average only 50 per cent of the EU citizens had an optimistic view of biotechnology (EU 1997). The share of optimistic people was biggest in Finland compared to all other EU countries, a fact that can be seen as an important supportive environmental factor which may open up great prospects to biotechnology in Finland.

We must be cautious, however, not to become too optimistic concerning the future of biotechnology in this country. A company survey conducted in 1997 shows that Finland, together with other Nordic countries, Austria, Switzerland, Germany and Greece, is considered as having a less supportive environment for investments in and use of biotechnology. The following elements were considered as the most important influencing factors: market conditions, consumer acceptance, intellectual property, regulatory framework, entrepreneurship, the science base, availability of equity capital, skilled staff, availability of suppliers and fiscal policy.

Constructing the Biotechnology Competence

Biotechnology is very much dependent on formal education, as - according to Lundvall - knowledge in this field is highly codified. This is reflected in the fact that Finland has certainly invested heavily in biotechnological education in the last 15 years. Since the mid 1980s, bio-engineers have been educated at the Helsinki University of Technology. In 1998, a master's programme in biotechnology was established at the University of Turku. In 2000, sixty new students started their studies in a new training programme in biotechnology at the University of Kuopio.

In the Science Park of Viikki in Helsinki, about 3,000 students study different subjects of biotechnology and life sciences. Finnish polytechnics too offer training programmes in biotechnology. However, the number of biotech students is not yet comparable with the number of engineering students, for example. On the university level, about 500 students start their studies in biotechnology each year, while about 3,500 start their engineering education.

Furthermore, fourteen graduate schools have been established at Finnish universities supporting research in various areas of life sciences. Five of them are situated in Helsinki, four in Turku, two in Kuopio and one in Oulu, Tampere and Joensuu respectively. The location of graduate schools reveals something of the anticipated location of Finnish biotechnology industries; the major regions of the future biotech industries seem to be the Helsinki and Turku regions. The graduate schools are mainly integrated in regional biotechnology centres of expertise.

The Pharmaceutical Industry

In Finnish industrial policy discourse, biotechnology is mainly associated with the pharmaceutical industry. The leading role of the pharmaceutical industry can be explained to some extent by the fact that the workforce in this sector is traditionally highly skilled with many academics, having a degree in medicine or technical sciences, working in R&D departments. The fact that private ownership dominates in this industry while it does not in other biotechnology fields may be seen as an important factor as well. Public opinion is also supportive; because of the careful clinical testing system and the regulations in medical industries, people accept biotechnological solutions in drugs more easily than in other product branches.

The crisis of the pharmaceutical industry at the beginning of the 1990s may also have contributed to the dominance of medicine in Finnish biotechnology. Many companies in the industry changed their strategies and abandoned their biotechnological research and development activities. As a consequence of this, many highly educated employees in companies' R&D laboratories lost their jobs and began to start their own small R&D firms. The founding of the new centres of expertise also strengthened the development; in these new centres there was a proper support structure available for small R&D enterprises.

Today the new university-affiliated biocentres - or the Centres of Biotechnological Expertise - have an important role to play in the research and development activities of the pharmaceutical industry. The BioCity centre in Turku, Biocenter in Oulu and the A.I. Virtanen Institute in Kuopio all work in the field of medical and diagnostical research. They have strong links to pharmaceutical industries, but they have also been criticised for excessive use of limited resources. These Centres of Expertise take in a lot of small firms, spin-offs either from universities or larger pharmaceutical companies.

The founding of the Centres of Biotechnological Expertise has changed the structure of the Finnish pharmaceutical industry significantly. Previously, only a few large national companies having their own research units existed. Now a network of small R&D-based firms situated in the university-affiliated Biotech Centres of Expertise has become a key actor in this biotechnology sector. These enterprises can be divided into three different types.

Firms belonging to the first type concentrate on research and development activities, while the second group of companies specialises in knowledge-intensive business services, and the third type of companies can be characterised as technology support-producing enterprises. Some global pharmaceutical companies also have smaller units in Finland and in Finnish biocentres. In addition to enterprises, the Centres of Expertise also host university institutes and private research institutes.

The Turku area in the south-west of Finland has attracted a remarkable number of Finnish pharmaceutical companies. To further develop this industrial agglomeration, the city of Turku has made large investments in modern biotechnology. The reason for the city's heavy investment in biotechnology is that it did not participate in the rapid development of the Finnish information and telecommunication technology. Finnish pharmaceutical corporations are small, however, which makes it difficult for them to hold their own in the global markets. Because of the limited scope of the national pharmaceutical industry, most R&D-oriented companies in Finland look for co-operation with global corporations. After the economic crisis at the beginning of the 1990s, the Finnish pharmaceutical industry not only restructured, but a number of mergers and take-overs took place, and some companies went bankrupt.

Nowadays one of the leading pharmaceutical corporations in Finland is Orion. Two of its divisions, Orion Pharma and Orion Diagnostica, concentrate on products based on biotechnological research and development work. Their net sales in 1998 amounted to about FIM 2,600 million, about half of Orion Corporation's total net sales.

Orion Pharma is engaged in the research, development, production and marketing of pharmaceuticals and drug substances, as well as drug-related products for the prevention and treatment of diseases. Their product portfolio encompasses all elementary therapy groups, including veterinary drugs. International operations account for approximately half of the total net sales. The growth of the business is increasingly based on drugs resulting from the company's own pharmaceutical research activity. Orion has several drugs manufacturing plants in Finland, and one in Denmark.

The focus of Orion's pharmaceutical research is on neurology, asthma, menopausal disorders, cancer and heart failure. In special fields, R&D activity comprises all phases of pharmaceutical research, from basic research in Finland to multinational clinical trials. R&D expenditure amounts to about 16 per cent of net sales of pharmaceutical reparations, and around 730 persons are engaged in R&D activities. Orion Pharma collaborates with bio-centres in Kuopio, Turku and Helsinki.

Orion Diagnostica is a division relying on biomedical research in its tests for diagnosing diseases. Its core competence is in infectious disease diagnostics, specific protein and hormone assays, and bone markers. International markets account for almost 80 per cent of the operations, most importantly Scandinavia, Central Europe, the USA and Japan. The division also has subsidiaries in the Nordic countries.

Orion Diagnostica cooperates with scientific institutes and company units in Medipolis in Oulu.

The Agro-food Industry

Conservatism and less favourable attitudes towards technological development are the main reasons for the low innovation capacity of the Finnish agro-food industry in the field of biotechnology. Moreover, exaggerated expectations concerning return on investment, the fear of technological risk and the lack of skills and competence in modern biotechnology can be seen as significant factors that have hampered biotechnology-related R&D activities in the Finnish agro-food industry.

The delayed social and economical transformation in Finland contributed to the high status of farmers in Finland. In Finnish agro-food industry, the farmers, their unions and their co-operatives represent the only relevant group that has an influence on techno-economic development. Most of the agro-food industrial mills, creameries, slaughterhouses and food factories have been owned by the farmers' production cooperatives. The share of privately owned companies in the food industry is very limited and they are rather small by size.

The co-operatives have become national monopolies. The co-operative creamery Valio, for example, was in possession of 97 per cent of the milk markets by the end of the 1970s. Correspondingly, the co-operative's market share in the meat industry was 53 per cent and its share in the egg industry was about one-third. This 'monopolar' structure in the Finnish agro-food industry, which prevented hard competition, has not been a supportive environment for any kind of innovative activities. Also, the fact that the industry was heavily subsidised became a hindrance for technological progress.

Before the Finnish EU membership and the opening of the markets in the 1990s, the structure of the Finnish agro-food industry can be described by the term 'agro-food industrial complex'. This 'agro-food industrial complex' is an economic system consisting of the industries providing agriculture with material and technical equipment, the industries taking care of the agro-food raw material production, and to the industries taking care of the processing and delivery of the products. In this tightly coupled structure, the state and its financing, research and educational organisations are entwined with the farmers' organisations and co-operatives. Also, privately owned companies had to adapt themselves to the structure and strategies of the agro-food industrial complex. The fact that the whole education system for the agricultural sector was subordinated to the Ministry of Agriculture until the end of the 1960s is a further indicator of the closeness of the sector.

The co-operative Valio founded at the beginning of the twentieth century by co-operative dairies to export butter plays an important role in the Finnish agro-food industry. Today, Valio is engaged in processing and marketing milk, dairy products and other food products over a total product range of 800 items. During the 1980s and 1990s Valio invested heavily in the research and development of functional food products. The company has been marketing products with Lactobacillus GG (LGG) in Finland under the Gefilus brand since 1990. Gefilus products include yoghurts, cultured buttermilks, pasteurised milk, a fermented whey drink, fruit drinks, as well as capsules. Today, products containing LGG are on the market in more than twenty countries around the world, under licence agreements from Valio.

The company is also the world leader in applying ultra-high temperature (UHT) technology to infant foods. The company has developed (UHT) technology in close co-operation with paediatricians. Several clinical studies have been carried out in order to improve the composition of its products. For nearly 20 years, Valio has also studied, experimented with and put into practice lactose hydrolysis in a highly sophisticated and intensive manner.

The company has developed tailor-made and patented solutions for the removal of lactose in different milk products. These solutions include the use of many soluble enzymes: the Valio Hydrolysis Process, using immobilised enzyme technology, and the Valio Chromatographic Process, in which lactose is physically separated from milk. Valio Ltd is owned by eleven co-operative dairies and 17,000 dairy farms produce milk for the company. The Valio Group comprises the parent company Valio Ltd and its subsidiaries in Finland and abroad. The Group employs 4,300 people.

The whole Valio Group net sales in 1998 amounted to FIM 7.6 billion (a1.3 billion), of which exports and foreign subsidiaries made up one-third. Noteworthy subsidiaries are Valio Sverige AB in Sweden and Finlandia Cheese Inc. in the USA, for example. Besides Valio's products, there are also other examples of successful R&D activities in the Finnish agro-food industry. Beer is a product with a long tradition in Finland. Lately malt-houses have focused on developing malting barley and new malting technologies.

The most recent addition to barley breeding techniques uses genetic engineering. The first genetic barley in the world was made in 1993 by the research group of the Technical Research Centre of Finland (VTT) and was used in trials. The best-known case is the development of the Benecol plant stanol ester, an ingredient that lowers serum cholesterol levels. The problem that triggered the innovation was to find an application for the surplus sitosterol of the chemical pulp industry of Kaukas Ltd. Raisio Ltd developed the idea of using the surplus sitosterol in margarine and vegetable oil. In the so-called North Karelia project, researchers discovered the cholesterol-lowering character of Benecol.

Nevertheless, Benecol, an innovation with great economic potential, demonstrates many of the problems related to the long-term economic isolation of Finnish agro-food industries. Raisio Ltd had great problems in getting access to the US market, as it lacked the knowledge of the rules and procedures regulating the functional food market in the USA. Due to the lack of expertise, competitors were able to bypass the R&D work conducted on Benecol and develop their own products of a similar kind.

The Raisio Group was established by Finnish farmers in 1939. The first production unit was a flour mill, but the company has over the years expanded into many other areas of arable product processing, and it can claim to be ahead in this field of knowledge. The Raisio Group headquarters are in Raisio, a small town in the south-west of Finland. The Group employs nearly 2,900 people, of which about 40 per cent are based outside Finland. Today Raisio has production units at over forty locations in seventeen countries. The Group's turnover was FIM 4,950 million (€833 million) in 1998. The parent company, Raisio Group plc, has been listed on Helsinki Stock Exchange since 1989, and over 50 per cent of its shares are owned by international institutional investors and funds.

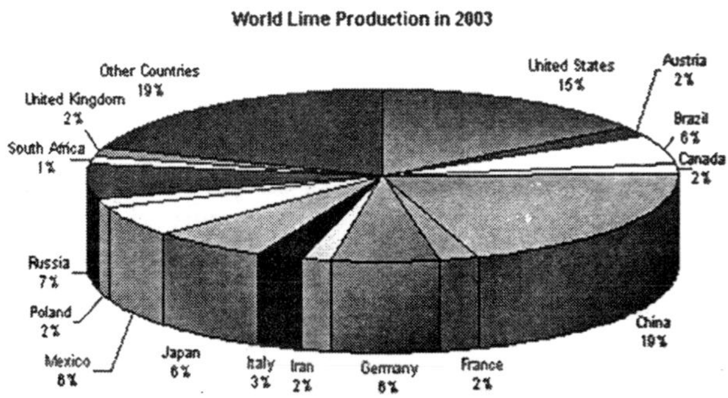

Fig. World lime production

The Raisio Group exports margarine products, malt products, wheat flour, pastas, cereal flakes, potato products, cat and dog food, feeds for fur animals and farm feeds. The Chemical Division is Raisio's most international division, with production plants in fifteen countries. It is a specialist supplier to the pulp and paper industry throughout the world. Its exceptionally wide product range includes chemicals and equipment extending from pulp-making to the manufacture, coating, conversion and recycling of paper.

After the opening of markets and Finland's EU membership, the Finnish agro-food industry has approached the situation of escalating change. One example of the fundamental nature of the changes is the course of events around Cultor Ltd. The corporation employs 7,000 people in Finland. Cultor is the world's largest producer of functional ingredients for the international food industry and the world's leading supplier of feed ingredients.

The company is also the leading producer of emulsifiers, functional systems, locust bean gum, fat replacers and butter flavourings, and is the world's second largest producer of pectin. The product range also includes other textural ingredients, flavourings, enzymes, sweeteners, antioxidants, and other protectants, in which Cultor is among the world's leading suppliers. Cultor has invested a lot in biotechnology-oriented R&D activities.

The development of the sweetener xylitol is one success story in applying biotechnology in the Finnish agro-food industry and Cultor. The research done in the Faculty of Dental Surgery in the University of Turku addressed the beneficial effects of xylitol for human teeth. Cultor Ltd placed xylitol in many of its products, which led to economic successes. In 1992, the R&D and the production of xylitol received the innovation reward of the Finnish National Fund for Research and Development. In March 1999, the company accepted the bid of the Danish corporation Danisco. In November 1999, Danisco was listed as the first non-Finnish company on the HEX Helsinki Stock Exchanges' Main List of blue-chip companies.

Cultor continues to operate as Danisco Cultor in the ingredients sector of the Danisco concern. Most of its products are produced using natural raw materials, such as vegetable oil, seaweed, citrus fruits and seed from leguminous plants. The Danisco strategy is to establish production plants close to key raw materials. Danisco is also developing and testing the GMO sugar beet in collaboration with DLF-Trifolium A/S and Monsanto Company. The case of Cultor illustrates that even bigger Finnish companies in the biotechnology sector have difficulties in holding their own in a globalising market.

This may be due to the fact that, in biotechnology, huge investments in R&D have to be made to be able to compete globally. Nonetheless, Danisco has cut the R&D activities in

Cultor, and at least 15-20 per cent of the researchers have been made redundant. Recently, the entire agro-food sector in Finland has been undergoing major changes, turning it into a research-based modern biotechnology industry.

The Agricultural Research Centre of Finland is planning the experimental production of genetically modified cheese and it has started research on GMO animals. According to the OECD's database of biotechnological field trials, two projects researching the streptococcus bacterium are being conducted in the Valio laboratories. Kemira Agro Ltd is studying the virus resistance of potatoes and Hilleshog AB is studying the glyphosate tolerance of sugar beet. In total there were six biotechnological field trials conducted in Finland before 1997.

Other key actors in the agro-food biotechnological R&D field, besides the laboratories of the agro-food industry and the Research Centre for Finnish Agriculture, are the new Helsinki Science Park in Viikki, which employs about 1,000 researchers, and the Kuopio Technology Centre TEKNIA Ltd. These new agro-food biotechnology centres of expertise represent the new technology policy of Finland. The new centres of expertise establish broader and closer linkages between universities and industries. Education, university research, knowledge-based agro-food industrial small firms, and departments of larger companies all are operating in close connection with each other in these centres.

The only Faculty of Agriculture and Forestry in Finland is now established in Helsinki Science Park. In this park, other research institutes are also operating: the Institute of Biochemistry, the Institute of Microbiology, the Institute for Animal Physiology, the Institute for Environmental Research and the Institute for Pharmacy, for example. The Helsinki Science Park is planned to become the most extensive concentration of biosciences in Europe.

In 1995, the Development Company Foodwest in Seinajoki was launched. It is owned by twenty-four agro-food companies and thirty-two communes and communities, among them the Central Union of Agricultural Producers and

Forest Owners (MTK). A few years ago, Foodwest was elected a Centre of Expertise, and today it employs about fifteen people. There is also a Development Company for agriculture production called Agropolis Ltd in Jokioinen. It is owned by local communities and other public authorities.

The Wood-processing Industries

In Finland, the production and export of chemical pulp - ultimately a semi-manufactured product and raw material to the paper industry - has been comparatively high. The manufacturing of chemical pulp is a representative example of mass production. For a long time, one problem in the Finnish wood-processing industry was the low degree of horizontal process integration. This is a result of the Finnish so-called smallholding policy; two-thirds of Finnish forests are nowadays privately owned by farmers or testamentary heirs of ex-farmers, one-quarter belongs to the state, and only 8 per cent to corporations.

Due to this situation, a kind of dichotomy in the production chain of the wood-processing industry emerged, which has drawn a clear dividing line between corporations on one side and farmers on the other. Both interest groups formed their own economic and commercial organisations. In the 1960s, the situation had already developed into a kind of 'double-closure' and opposing national cartels. Not surprisingly, the research and development interests of the two key actors also diverged into two different directions.

The wood-processing industry was mainly interested in research and development of end products and processes inside the factories. Correspondingly, the farmers' (or 'wood growers', as they are called) interest focused on productivity improvement of forests. In general, biotechnologically oriented research made a larger impact on the wood growers' side, and less so on the corporations' side. However, the wood growers were not greatly interested in intensifying forestry yield through biotechnological applications. Because of the fact that Finland is rich in forests, the wood growers were mainly interested in the improvement of forestry work.

The productivity of Finnish forests was defined very early as an issue of national survival. The state has - beside the Finnish wood grower organisations - been active in forestry research and development. The great interest of the state in the wood-processing industry is self-evident, as some 40 years ago 70 per cent of Finnish exports came from this industry. Due to the polarised situation in the Finnish wood-processing industry, the state took on a kind of umbrella role. It established a broader vision in which both the interests of the corporations and of the farmers can be taken into account.

Today, there are three globally acting corporations in the Finnish forest industry. The UPM-Kymmene Group is, measured by turnover, the biggest; Stora Enso is one of the world's leading forest industry companies; and Metsa-Serla Oyj is the sixth largest forest industry and paper-producing company in Europe. The large forest industry companies, however, seem to be less active in promoting biotechnology innovations, while companies in the chemical pulp industries, being part of the forest cluster, seem to be much more engaged in research in modern biotechnology.

The most famous biotechnological innovation in the chemical pulp industry, developed by the Technical Research Centre of Finland VTT's biotechnological laboratory, is an enzyme-aided process for bleaching cellulose pulp. In this process, hemicellulases are used in order to avoid using chlorine chemicals. The first pulp mill designed for complete chlorine-free bleaching was built at OY MetsaRauma Ab and started in 1996. Ahlstrom is the leading company in Finland in developing new processes and machinery for pulp and paper manufacture in which the environmental aspect is considered.

The structure of the biotechnology innovation network in forest industries differs significantly from the one in the pharmaceutical industries. There are different research organisations functioning in the beginning and end of the production chain. The chemical pulp and paper companies and the Central Laboratory KCL co-operate in research and development of enzymes with companies like Raisio Chemicals, Primalko and Genencor International. The latter co-operate with VTT Biotechnology, which has close links to university research. In the beginning of the production chain, there are actors working on improvements in the productivity of forests, such as Forest Centres and The Finnish Forest Research Institute (METLA).

The Finnish Forest Research Institute, established in 1917, is an independent research organisation subordinated to the Ministry of Agriculture and Forestry. This central forest research organisation has a staff of 700 people, 200 of whom are research officers in the research centres located in Helsinki and Vantaa. METLA also has eight research units in different parts of the country. METLA has a strong tradition in research and development activities based on traditional biotechnology. Applications of modern biotechnology have often been delayed, mainly because of the problematic proprietorship of Finnish forests. Wood production based on applications of modern biotechnology would be easier to realise in a large and market-oriented forest industry than under the conditions of dispersed forest smallholding. Despite these difficulties, six modern biotechnological, forest genetic research projects are currently conducted at METLA.

Concerning corporations, the most interesting institutional actor has been the previously mentioned Central Laboratory KCL, which concentrates on quality control of the end products of chemical pulp and paper industry mills. But lately the interest of the laboratory has broadened from mere quality control to larger R&D activities, including also research in biotechnology. The state research organisation, the Technical Research Centre of Finland (VTT) with its biotechnological laboratory is another important actor that has close links to corporations. Although most of the biotechnology projects of METLA, VTT and other institutional actors are financed by the WoodWisdom programme, modern biotechnology has a marginal role in the whole programme entity.

The radical changes at the beginning of the 1990s broke with the customary notion of the juxtaposition of wood-processing corporations and wood grower organisations. In the new atmosphere, the wood-processing industries began to be considered as an uninterrupted production chain from forestry to consumption. According to the Technology Development Centre of Finland, the greatest future challenge for the wood-processing industry is 'to integrate forest economy and forest research to a constant part of the research of products of wood-processing industries. This means comprehension and governance of the linkages between the wood raw material quality and the end product quality'. Research in the Bio and Food Technology Unit of the Technical Research Centre of Finland is in this direction; it applies biotechnology in the wood-processing industry in several research and development projects. The research on wood material, for example, aims at increasing biological stability and hindering the dyeing of the raw material.

Biotechnology Supporting Policies

The Finnish Ministry of Education established the first national research programme on biotechnology in 1987. The aim of the programme was to develop four powerful Centres of Biotechnological Expertise by 1992. These new centres were planned to affiliate with the Finnish universities that were assessed as having the capacity and resources to develop the new field of scientific research. The Centres of Biotechnological Expertise were set up in Helsinki, Turku, Kuopio and Oulu. By the end of the 5-year period, the Ministry of Education made an evaluation aiming at identifying needs to further develop biotechnology and molecular biology. The Ministry decided to continue the programme until the end of 1996. The programme period was later extended to 2000.

Besides the Ministry of Education, the Ministries of Trade and Industry, Agriculture and Forestry, and Social Affairs and Health took part in the financing of the biotechnological research programme. The four Centres of Biotechnological Expertise became part of the Finnish network of 'Centres of Expertise'. Nowadays this network consists of seventeen Centres of Expertise, of which seven are large and have an established status. Four centres have been working for some time and six are new and rather small ones. The new Centres of Expertise in Seinajoki, Lahti and Pori are located in regions without a local university but they are linked to large regional polytechnics or local annexes of universities.

Basic and Applied Research

Research and development in biotechnology is now conducted in the centres of Tampere and Seinajoki in addition to the traditional centres. In Finland, the Centres of Expertise Programme is intended to become the core of a newly developing regional policy. The strong relationship between the state's regional financing and the Centres of Expertise Programme has led to 'excess demand' for new centres. Therefore the policy was criticised for the excessive decentralisation of limited R&D resources. The result is that, in the field of biotechnology, no new Centre of Expertise will be set up in the near future.

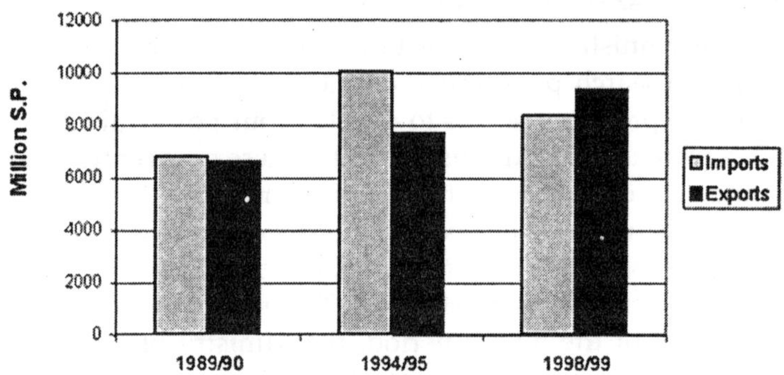

Fig. Agricultural and agro-food trade report

Centres of Expertise offer local companies business services, and forge links between regional enterprises and local universities or - in some cases - local polytechnics. They are presumed to be sources of knowledge for regional economies. They also have an important role to play in the transfer of technology from high-tech centres in Finland and abroad to their own region. In addition, they are expected to generate new forms of business and to support the foundation of new enterprises, as well as to create a stimulating environment for the techno-organisational modernisation of the existing companies.

The university - according to the basic philosophy of the Centre of Expertise Programme - can stimulate and support regional development in high-tech sectors. In case there is no university located close to the Centre of Expertise, a regional polytechnic may, as the only base and core of local industrial network, begin to take on the university's function.

The fact that the new Centres of Expertise will be located in towns without a local university may contribute to an academic drift to polytechnics. This phenomenon, found by Pratt and Burgess, has contributed to the fact that, by the end of the 1980s, the whole polytechnic system was dismantled in the UK and the polytechnics were transformed into universities. In regions without a local university, polytechnics may well become the base and core of local industrial networks, as was the case in the UK.

Four Centres of Biotechnological Expertise, Oulu, Turku, Tampere and Kuopio, focus on medical research and co-operation with pharmaceutical industry. The focal area of Helsinki Science Park and Foodwest in Seinajoki is the co-operation with the agro-food industry. The dominant status of the pharmaceutical industry in the field of Finnish biotechnology can partly be explained by the fact that this was the focus area of the old Centres of Expertise. As those centres that concentrate on R&D in the agro-food field have only recently been established, this area is still lacking behind in its economic development.

Strong Public Financing

Research and development in biotechnology is highly dependent on state funding. As has been mentioned earlier, only 0.4 per cent of the private venture capital investments was directed to biotechnology in 1997. As a capital-needy country, state and public financing plays an important role in Finnish venture capital markets. Different parts of the idea-innovation chain are financed by different institutions. The

Academy of Finland is supporting basic scientific research in universities and research centres.

The Technology Development Centre aims at applied research and product development by supporting the co-operative activities between university research and companies. The role of the Finnish National Fund for Research and Development (Sitra) is to support promising start-up enterprises. Later on, the new knowledge-intensive companies are expected to get their financing from the capital markets.

Tekes invested FIM 535 million (€90 million) in the field of chemistry and biotechnology, which is 27 per cent of its total investments. The Academy of Finland and Tekes invest about FIM 20 million (€3.4 million) a year in research in biosciences and biotechnology as part of the Centres of Excellence Programme. In addition, Biocentrum Helsinki, Biocenter Turku and BioCity Turku are supported with FIM 4 million (€0.7 million). This means that 40 per cent of the aggregated financial support of the Academy of Finland and Tekes is directed to biotechnology. The national programme for the research on biotechnology, started in 1988, invests about FIM 80 million (€13.5 million) a year. As state financing of Finnish biotechnology has been quite large in the 1990s, some criticism has been made about the results of this massive investment.

Presently, there are two research programmes going on within the common foundation of Tekes and the Academy of Finland. The Genome Research Programme includes gene regulation. It studies interactions between several genes and gene products, gene transfers, and knock-outs as well as gene therapy. The budget of the programme amounts to FIM 89 million (€15 million) in 6 years. The Academy of Finland financed twenty projects with FIM 37 million (€6.1 million) from 1995 to 1997. From 1998 to 2000, the Academy of Finland is financing fifteen projects with FIM 24 million (€4 million). Tekes financed five projects with FIM 5 million (€0.8 million) in 1998.

The Cell Biology Research Programme studies mechanisms of cell division and differentiation, the biogenesis of cell organelles and intracellular trafficking as well as signal transduction. The programme started in 1998 and will continue until 2001. The budget of the programme amounts to FIM 32 million (€5.4 million). The Academy of Finland is financing nineteen projects with FIM 22 million (€3.6 million). Tekes financed four projects with FIM 5 million from 1998 to 1999.

In 1993, the Ministry of Education established a new centres of excellence policy by ordering the Finnish Higher Education Evaluation Council to choose ten units and institutes as 'top units'. This policy of establishing Centres of Excellence has become the core of the Academy of Finland's policy. It expects that the new policy can lead to a strong concentration of research and innovative power. Currently, 6 per cent of the Academy's funding is directed to the Centres of Excellence Programme. In 2000, a total of twenty-six new Centres of Excellence will be established. Nine of them are operating in the field of modern biotechnology and biosciences.

The Academy of Finland also financed research in modern biotechnology within the framework of its basic research programme. There is, for example, an ongoing programme for gene research and cell biology. In 1999, the Academy of Finland started the Research Programme for the Biology of Structures and, in the following year, the Research Programme for Biological Functions.

The Academy of Finland is financing the Research Programme in Molecular Epidemiology and Evolution. The primary areas of this programme are population genetics, concerned particularly with the evolution of genes, the adaptation of organisms to extreme conditions, genetic epidemiology and the Finnish genetic heritage, and environmental molecular genetics, including genetic factors predisposing to disease.

Tekes started a 5-year Biotechnology Development Programme in 1988. It consisted of five subjects important to Finnish industry: biotechnology for the pulp and paper industries, bio-process engineering, plant biotechnology, animal cell technology and biologically active molecules. Tekes also has an agro-food biotech programme named 'The Innovation in Foods Programme'. It aims at supporting the production of more competitive foods through research. The goal of the programme is to ensure the production of increasingly high-quality products by improving the standards of food technology and related research in Finland. The programme started in 1997 and continued until 2000. The budget of the programme will be over FIM 200 million (€33.7 million) during 4 years. All the significant Finnish agro-food companies are participating in the programme.

In 1991, the Finnish National Fund for Research and Development (Sitra), aimed at financially supporting new start-ups, invested FIM 400,000 (€67,000) in biotechnology, which was only 1.2 per cent of its total investments. Eight years later, the investments in biotechnology amounted to FIM 35.7 million (€6 million). Altogether Sitra has invested in eleven biotech companies; today, the share of biotech investments is already 9.5 per cent of all investments. In 1997, two special funds, Sitra Bioventures Ky and Sitra Bio Fund Management Ltd, were started with a capital of FIM 150 million (€25.2 million).

Biotechnology industries have only played a minor role in the Finnish industry up to now. However, the state and state-owned institutes have started to invest heavily in biotechnological research projects, as well as in education and research infrastructure, as part of an extension policy of Finnish export's industrial basis. We are entitled to use the word 'massive' because only little R&D so far comes from the private sector, and only few business firms have started production of biotechnology-based products.

The structure of the Finnish biotechnology sector is similar to that of European biotechnology. The pharmaceutical industry represents the leading sector, while the agro-food sector and the forest industry are lacking behind partly due to the specific 'lock-ins' described in this chapter. But there are signs that the industries will be able to overcome the problems caused by the lock-ins.

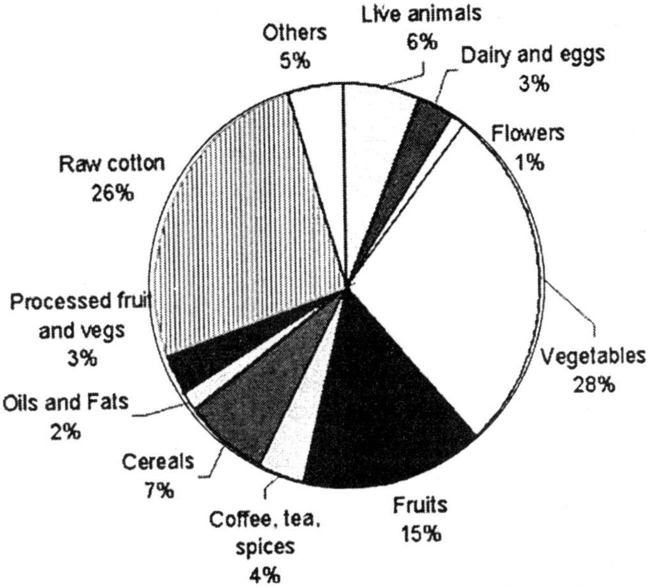

Fig. Composition of agricultural exports to the world

The further development of the biotechnology sector in Finland depends on whether the emerging research infrastructure will soon lead to the building up of a significant number of manufacturing plants. In overall terms, the business environment in Finland seems to be less supportive for investment in and use of biotechnology than in the USA and some other parts of Europe. However, there are also some supportive factors. There seems to be widespread agreement that biotechnology could become the fourth pillar of the Finnish export sector, indicated by the massive public investment in research. Furthermore, the Finns in general are very open to technological development and their attitudes to biotechnology seem to be less negative than in other European countries. And last but not least, Finnish technical universities are known for their openness to close cooperation with industry. The Centres of Expertise seem to create a supportive environment for close university-industry co-operation.

There are also important factors that can hinder the further development of the biotechnology sector in Finland. First, there seems to be a serious shortage of highly qualified scientific staff in the sector, which the government is aiming to overcome by heavy investments in the education infrastructure. There is, however, huge demand for qualified engineers in the IT sector as well. Whether in the competition over human resources the biotechnology sector will be able to hold its own remains to be seen. Anyhow, there is a lack of students oriented in mathematics and natural sciences in their matriculation examination.

A major disadvantage is the small size of the Finnish market. Therefore, companies in Finland have to be globally oriented from the very beginning. However, as some examples demonstrate, the small Finnish biotechnology firms, particularly in the agro-food and pharmaceutical sector, often lack the management capacity to establish themselves in foreign and particularly in the most important US market. Companies operating on the national or even on the European level may soon face serious problems because they lack sufficient awareness of and access to new biotechnological knowledge.

Concerning regulatory aspects, Finnish firms suffer the same fate as other European firms. The European regulatory framework seems to have a negative effect on the competitiveness of European firms. The greatest losers in the biotechnology age are likely to be the farmers who, for regulatory reasons, have been denied access to some products. As European farmers in general, the Finnish farmers are likely to lose global market shares to farmers from other parts of the world, who are already adopting products using biotechnology.

Still another problem is the small size of Finnish firms, particularly in the agro-food and pharmaceutical industry. Small companies face serious problems as product innovation and improvement in operating efficiency come to depend increasingly on the use of biotechnology. In order to be able to develop the absorptive capacity necessary for applying biotechnological knowledge, companies have to set up respective research capacity.

This is a major problem for small companies, as can also be demonstrated by the small amount of private R&D investment of Finnish firms in biotechnology. However, Finnish firms are starting to co-operate and merge not only with European but also with US companies to overcome the disadvantage of being small. As the pulp and paper industry includes some global players, the sector is not affected by this problem. As some huge mergers have already taken place in this sector, Finnish pulp and paper companies may well be able to retain their place in the biotechnology era.

Chapter 8

The Competencies of Regions

CLUSTERS IN BIOTECHNOLOGY

Comparative analysis suggests that biotechnology tends to cluster regionally. In both the United States and the United Kingdom, a few states or metropolitan areas concentrate the majority of firms. Factors leading to this strong clustering include access to university and public sector research, to venture capital and large markets in major cities. American biotechnology clusters developed around the US science base. Strong clusters exist in California, Maryland and Washington, DC, Massachusetts (Boston), New Jersey (Princeton), New York City, North Carolina (the Triangle Research Park), Pennsylvania and Texas.

These agglomerations undertake almost two-thirds of US biotechnology activity. Conversely, British clusters are closer to industrial agglomerations and concentrations of graduates. Analysts also consider the hypothesis that in Britain, the size of the biotechnology cluster varies with population: the triangle region delimited by London, Oxford and Cambridge has 50 per cent of British dedicated biotechnology firms (DBFs). Greater London has the biggest agglomeration. However, no variable was highly significant in the United Kingdom to explain biotechnology clustering.

Knowledge spill-overs, or externalities, are responsible for the geographical clustering of biotechnology firms. Innovative organisations generate knowledge, some of which 'leaks' towards other organisations. As much of this knowledge is tacit - particularly in new activities such as biotechnology - geographical proximity is the key factor for all organisations absorbing such externalities.

However, most of the neoclassical literature in economics assumes that all organisations (firms, government laboratories and universities) are equally innovative. We challenge this assumption and inquire about their different capabilities in this area. We also analyse the relationship between those capabilities and the clustering success of entry attractors. For this purpose, we extend the discussion of competencies from business administration to cluster theory.

THEORY OF THE COMPETENCIES OF REGIONS

The resource-based theory of the firm explains the different performance of firms by their different endowment of factors. Some firms develop internally or secure externally these resources to become more competitive than others in a specific field or industry. Thus some firms get superior rents and maintain these idiosyncratic advantages for some time. A differential performance may arise both from divergent endowment of resources, and from their novel combination of resources. The competence theory of the firm thus goes one step beyond the resource-based theory, recognising that superior routines and organisation may yield better performance.

Competitors do not immediately imitate these performances because routines can not be observed easily, and because there is always some level of ambiguity about the causes of this superior attainment. The competence theory for regions has identical assumptions for firms concerning this point. Like firms, regions may be understood as sets of competencies and resources both tangible (physical infrastructure, corporate physical assets, research and development [R&D] laboratories) and intangible (skills and human capital, routines, institutions). Not only resources are important for regions, but also the capacity to use them to produce superior results.

In the evolutionary theory of the firm, as well as in the capabilities and competence approach, 'organisational forms do matter because information flows and behaviours differ according to the particular "institutional architecture" of each system. In particular, if each system's performance rests on specific learning dynamics by individuals or groups (such as firms), the institutional architecture affects the scope in which such learning can occur'. Universities, of course, enhance regional capabilities by producing knowledge that facilitates wealth creation and the quality of life.

Similarly, regions within nations, like firms within industries, differ persistently in resources, behaviour and performance. Thus their different outcomes and growth rates may be affected by some hidden variable, such as competencies, particularly core competencies. These variations in outcome suggest unequal resource endowments particularly in skills, information and preferences, different attitudes towards risk and contrasting institutions related to technology and innovation (such as government laboratories and universities), learning routines and financial institutions (such as venture capital firms).

Core competencies of regions include those that create economic value for markets outside the region, that are co-specialised, and difficult to imitate. In traditional industrial districts, as well as in the Perroux poles, regional competencies consist of, first and foremost, those of the region's firms, including their capacity to co-operate with each other. Core competencies usually transcend those of firms: in science-based agglomerations and technological districts, where SBIs prosper, they include the propensity and capacity to co-operate with and learn from other institutions in the regional system of innovation (RSI), such as local universities, government laboratories and venture capital firms.

Competencies, like knowledge in general, increase with use and decline if they are not used. Being closely related to and made of knowledge, they increase with practice, usually procuring sustained advantage to regions as well as firms.

Much knowledge is produced in an interactive way. Some authors have suggested that the density of the interactions between economic agents within a region also influences knowledge creation and thus performance. According to some authors, a region such as Alsace - where there is only little interaction among firms and with local universities and public laboratories, and where most firms are branch plants operating within an international network - does not constitute a regional system of innovation. The question of course remains open how much local co-operation and how many local spill-overs constitute a local or regional system of innovation.

As is the case in business organisations, the results for a region are not always evident. There is much causal ambiguity in the factors and/or their combination, that explain the performances of regions. Thus economic actors as well as governments act under assumptions of bounded rationality. In an opaque, murky environment, they invest and operate without precise knowledge of future results. We see below that some government laboratories and university research efforts generate few economic spillovers in certain regions.

Finally, competence theory goes beyond institutional analysis. Competencies point to skills, both personal and social or organisational. One may find the same institutions (universities and government laboratories for instance) in developed and developing countries, but endowed with totally different individual and organisational capabilities or competencies. Competence theory thus draws a link between human capital theories and institutional and evolutionary approaches.

CANADIAN BIOTECHNOLOGY

Canadian biotechnology started some years after the American competitors, but is comparable proportionately to the US field. By 1997, Canada had 282 domestic biotechnology DBFs, and close to 350 in 2000, seventy-one of them public. The total market capitalisation of these seventy-one firms was about C$12.9 billion by the end of May 1999, two-thirds of it in the four largest. Forty-two other firms had a market value under C$50 million each, and ten of these had less than C$10 million. Moreover, the window of opportunity seemed to be closing for biotechnology initial public offerings (IPOs): in the first 6 months of 1999 only one new firm entered the market, with an offer that collected C$22 million. Note that almost all of the largest firms are active in human health products.

Among the 282 DBFs existing in 1997, last year (in 1999), according to official figures, some 129 were active in human health (46 per cent), sixty-two in agricultural biotechnology, thirty-two in environment, twenty in food products and thirty-nine in all other areas. These 282 firms employed 9,800 people. Human health had over two-thirds of the employees, with an average size of eighty-one per firm, against forty-two in agricultural biotechnology, twenty-two in environment, twenty in food production and forty-three in other areas. DBFs had total annual revenues of C$11.2 billion, including exports amounting to C$4.9 billion.

The Canadian venture capital market has developed very rapidly in the last 10 years. By the end of 1998, the total pool was estimated at C$10 billion, and the same year more than C$1.66 billion was invested in some 1,200 companies, sixty of them active in biotechnology. Seed money was the most frequent type of investment in DBFs, the average investment being around C$1 million. With some 210 firms outside the capital market, it is apparent that many demands for venture capital have not been accepted. In other words, even if the Canadian venture capital market is sizeable, it cannot accommodate all the demands from domestic DBFs.

University research in health sciences concentrates in a few major metropolitan areas, such as Toronto, Montreal and Vancouver. In 1997, according to the latest figures available, metropolitan Toronto hosted 1,149 university professors in health science (at University of Toronto and McMaster University), in comparison to only 780 in Montreal (at McGill University, the University of Montreal and University of Quebec). Vancouver was third with 506 professors, all of them at the University of British Columbia. Calgary followed with

325 (University of Calgary), Edmonton with 305 (University of Alberta) and Quebec City (Laval University) with 249.

The Research Design

We based this chapter on an empirical analysis of technical competencies of DBFs, as indicated on the number of biotechnology patents granted in the United States to Canadian corporations, to Canadian government laboratories and to Canadian universities. Patents are uniquely revealing to study competencies, pointing not simply to resources (such as R&D personnel or laboratories), but to the ability to use resources in order to produce valuable results. Patents are thus the output of those resources, and represent a more useful indicator of competencies than simply R&D inputs. Due to a high propensity to patent in biotechnology, patents represent good indicators of competencies in this particular area.

In order to study technical competencies, we chose the US patent database instead of the Canadian one, for several reasons. First, all major inventions are patented in the United States as most Canadian companies, government laboratories and universities seek to protect intellectual property of their most valuable knowledge assets in the largest, most inventive and most affluent nation in the world.

Second, the US patent database, unlike the Canadian one, indicates the geographical location (country, city and state or province) where the invention has been made, thus allowing precise identification of the region and of the competence that has produced the invention. Third, more Canadian firms patent their novelties in the United States than in Canada. For instance, between 1980 and 1999 sixty-two Canadian biotechnology enterprises sought patents in the United States, and only thirty-seven in Canada.

Biotechnology presents a major methodological problem: it has no precise classification code - be it a patent code, an industrial (or SIC) code, or a trade code - because it is a 'genetic technology' used across a variety of industries. Thus we decided to identify DBFs (by means of the 1998 BIOTECanada

Biotechnology Directory). We examined the patents of these companies, and ascribed them to biotechnology, unless they were totally unrelated.

The presence of a set of keywords in the patents' names and abstracts decided whether the patent was biotechnological or not. These key words included terms such as 'DANN', 'monoclonal antibodies' and 'genetic material'. We decided that DBFs produce biotechnology patents, but we checked for supplementary proof of capabilities in the specific areas by examining patent descriptions.

An additional problem for studies using patents is the propensity to patent products rather than processes. Thus, we may have missed some process capabilities in biotechnology.

Government Laboratories

We examined government laboratories and all Canadian universities regarding their role in attracting or nurturing private firms' competency. We found six public laboratories (five with the National Research Council of Canada (NRC), and one provincial unit) with Canadian patents. A brief description of these laboratories is presented here.

The NRC launched five institutes for biotechnology in Canada after the passing of Canada's biotechnology strategy in 1983. Some of these institutes started from scratch while others benefited from the experience and assets of pre-existing NRC laboratories. These are the Biotechnology Research Institute (BRI), the Institute for Biodiagnostics (IBD), the Institute for Biological Sciences (IBS), the Institute for Marine Biosciences (IMB) and the Plant Biotechnology Institute (PBI).

The Biotechnology Research Institute of Montreal started from scratch in 1987 and now employs almost 200 scientists, plus another 200 guest researchers in pharmaceutical, environmental and bio-processing industries. The recently founded (1992) Institute for Biodiagnostics in Winnipeg is concentrated in developing technologies for disease diagnostic. The Institute for Biological Sciences in Ottawa, one of the oldest of the NRC (1916) focuses on immunochemistry.

The Institute for Marine Biosciences, originally established in 1952, works on marine biotechnology, including plants, shellfish toxins and natural chemistry. In 1948, NRC created an institute for plant biology in Saskatoon, known as the Prairie Region Laboratory; under the new strategy, it was revamped and became the Plant Biotechnology Institute in 1985, hosting some forty-five staff and seventy guest researchers.

Each institute has been an 'entry attractor' for new DBFs in its respective technology and region. As Canada had few DBFs in 1983, the NRC laboratories have antedated the founding of most of them. All these institutes, however, were created on or near university campuses, which already had departments or faculties of medicine, biology and biochemistry.

The Alberta Research Council (ARC) set up laboratories in Edmonton in 1936 and created biotechnology facilities in the late 1970s and the early 1980s. In terms of patents, ARC is the most active public laboratory in Canadian biotechnology.

Universities

Several Canadian universities have produced patents in biotechnology. The metropolitan areas of Montreal, Toronto, Saskatoon and Vancouver are among the most important centres of university research in biotechnology. We examined the competencies of a university via the patents held under its name, whatever the department or faculty having invented the product or process. Canadian universities have requested and obtained biotechnology patents since 1976, but most of them in the 1980s and 1990s. Research universities with a proven track record in medicine, plant and/or human biology include the Universities of British Columbia, Laval, McGill, McMaster, Montreal, Saskatchewan and Toronto.

The Evidence and its Significance

We collected patent data, as well as foundation dates, for the largest Canadian census metropolitan areas: Toronto, Montreal, Vancouver, Ottawa, Edmonton, Calgary, Winnipeg,

Halifax, Quebec City and Saskatoon. Only a few are hosts to public laboratories, but all have research universities.

The number of DBFs, of patents and patenting firms indicate that Toronto is the centre of Canada's biotechnology with nearly 50 per cent of the patents that have been granted to Canadian DBFs by the United States Patent Office since 1989. Despite the creation of the BRI, Montreal remains a distant second.

The two most inventive DBFs in terms of US patents are Toronto's Connaught and Allelix, followed by BioChem Pharma of Montreal. Two other Toronto firms come fourth and fifth: Visible Genetics and Spectral Diagnostics. Moreover, one major DBF dominates invention capability in Montreal (BioChem Pharma), and one each in Edmonton (Biomira), Vancouver (QLT (Quadra Logic Technologies) Phototherapeutics), and Saskatoon (Biostar). Each firm has between 38 and 75 per cent of the privately held patents in the region. Toronto appears more economically decentralised in terms of technological competencies in biotechnology: several firms own a large portfolio of patents.

As for public laboratories, IBS in Ottawa remains the most productive of the NRC's five (in terms of patents), but the Alberta Research Council in Edmonton is the most productive public inventor of biotechnology. Some of the public laboratories are not strong in terms of patented novelty and thus probably have less spillover potential in knowledge.

Concerning firm creation, BRI (Montreal) and PBI (Saskatoon) seem to have been major entry attractors: substantially more companies were created or attracted to the region after their creation. The other large NRC institutes in Ottawa, Halifax or Winnipeg did not have as much impact. Regions with major research universities, such as Toronto, Vancouver and Quebec City (the suburb Ste-Foy) seem to bring in many DBFs even without the help of government units. Research universities seem to be as important magnets as government laboratories, which possess some of the largest concentrations of R&D resources and competencies in the region where they are present.

The largest firms, measured by market capitalisation, are located almost all in Toronto, Montreal and Vancouver, with a few in Ottawa, Edmonton, Quebec City and London, Ontario. The dominance of the three majors in the capital market (Montreal's Biochem, Vancouver's QLT and Toronto's Biovail) is phenomenal. They represent nearly three-quarters of the total capitalisation of Canadian biotechnology.

University Laboratories

With regard to universities, the regions are different in rank. The University of British Columbia in Vancouver has most patents, followed by the University of Calgary, the University of Saskatchewan, McGill University and the University of Montreal (both in Montreal), and the University of Alberta in Edmonton. By urban agglomeration, Montreal and Saskatchewan share the first place in university patents, followed by Vancouver and Toronto.

Some agglomerations with university attractors did not bring in DBFs. Conversely, Canada's largest city, Toronto, has only one attractor (its university), but it is probably the most important centre for biotechnology in the country.

We thus proceeded to measure the precise weight of each factor. How important are public attractors (public laboratories and universities) to biotechnology clusters? Correlation analysis suggests that some of the government laboratories are of little importance. In fact, government laboratory capabilities measured by patents are negatively correlated with the three dependent variables: number of patents held by DBFs, number of DBFs and the number of patenting DBFs.

Conversely, university research strongly correlates with two of the three dependent variables: number of DBFs and number of patenting DBFs. University research thus seems to attract DBFs and to contribute to enhance their competencies. However, a university is not enough to entirely explain the location of firms.

The major overall attractor, however, is population. When we bring in the data from the 1996 Census Metropolitan Areas (CMAs), the three dependent variables strongly correlate with their total population. In other words, activity in biotechnology clusters where most people live, and secondarily, where Canadian universities do research. Population is at first sight a proxy for the size of the immediate market. Total population may also cover variables such as the size of the local pool of venture capital. (Unfortunately, we do not have figures in Canada showing the distribution of venture capital by cities.) The proportion of a CMA's population with university degrees is barely relevant. Only the size of the CMA and the scale of its university research matters.

Normalised correlation, holding population constant, confirms the significance of university research, at least for both total patents by DBFs and number of patenting DBFs. We calculated three ratios, dividing the three values of the dependent variables by the population for each CMA. Government patenting still has little impact - or even a negative one - on the dependent variables.

Finally, we conducted cluster analyses combining the patent productivity of DBFs, and of universities and government laboratories. The analysis uncovered three types of regions. The first type, consisting of Toronto, Montreal, Vancouver and Saskatoon, strong DBF and university patenting, combined with moderate government patenting. The second type, formed by Edmonton and Ottawa, has strong government patenting associated with weak DBF and academic patenting. The third group has weak DBF and public patenting, innovative dynamism is low. This group includes Calgary, Halifax, Kingston, Quebec City and Winnipeg.

It is to be noted that the regional pool of venture capital corroborates the previous ranking. Toronto, Montreal and Vancouver, in this order, are the centres of Canadian venture capital specialised in biotechnology. All the other regional agglomerations are far behind concerning the availability of venture capital funds for biotechnology. The university base, and the venture capital pool both reinforce the sheer effect of market size in the pecking order of Canada's clusters in biotechnology.

Policy Implications

In biotechnology, economic concentration of competencies goes hand in hand with regional concentration: only twelve firms hold over two-thirds of the patents. Six of them are in Toronto, two in Montreal, and one each in Edmonton, Ottawa, Saskatoon and Vancouver.

Toronto enjoys indisputable leadership as Canada's top biotechnology cluster, measured by the number of DBFs, the number of patenting DBFs and the patents. Montreal follows, Vancouver appears in the third rank. In other words, the three main Canadian biotechnology clusters are located in the three largest cities; all of them have at least one major research university and the largest venture capital pools.

With regard to theory, this analysis brings credibility to the competency theory of the region. Knowledge spillovers do not only depend on the (variable) amount of industrially useful knowledge produced by firms, universities and government laboratories, but also on the size of the immediate market and other related characteristics of the urban agglomeration. Relevant characteristics may include availability of venture capital (certainly much more readily obtainable in larger cities such as Montreal, Toronto and Vancouver than in smaller ones such as Edmonton or Saskatoon, whatever the quality of their university research and government facilities). 'Star scientists', as most researchers and university graduates in the health sciences, may also prefer larger cities.

Because of causal ambiguity, government laboratories have only marginally countered the market forces that tend to concentrate biotechnology activities in a few large cities with strong university and venture capital environments. If university creation represents a 'bottom-up' strategy and government laboratories a 'top-down' one, then the bottom-up approach to biotechnology seems adequate: universities are at the basis of most new firms and industrial dynamics.

This analysis suggests that governments might reinforce Canadian biotechnology clusters if they relocated their laboratories and research universities in the largest metropolitan areas, enjoying the strongest venture capital pools. Canada's citizenry and cities may not be large enough to bear the dispersion of efforts in ten or more urban metropolitan areas. In other words, government effort must concentrate where the potential for clustering is, thus in larger cities.

The size of the metropolitan area, the other major variable related to industrial clustering, may be a surrogate for other underlying variables, the existence of large hospitals (where clinical essays can be conducted) and a wider pool of human capital.

Chapter 9

Biotechnology and Biodiversity

THE LINKS

While the first section in this series of essays examined the positive and potentially less positive impacts of biotechnology on a range of activities in industrialised and developing countries, this section looks at the links between biotechnology and its key input, namely biological material. The revolutionary new products developed by biotechnologists depend greatly on the availability of germplasm and a knowledge of its properties and uses. It is the variability of this biological material that allows biotechnologists to examine and experiment with their properties to produce life-saving new drugs and higher yielding varieties of new crops.

Our knowledge of these characteristics and indeed of the total number of species that exist on earth is, however, limited. Nevertheless, the need to understand the manner in which species interact with one another and their environments has become more apparent with the widespread use of biotechnology.

This chapter examines the concept of biological diversity, its importance and our need to understand it, and the relationship between biotechnology and biological diversity. It begins by introducing the reader to the concept of biodiversity and the various definitions that have been given to it. The first section also discusses the reasons why biodiversity is acknowledged as an important resource for present and future generations, and why considerable difficulties exist in attaching appropriate values to this resource. This is followed by a brief overview of the literature on developing a framework for the conservation of biodiversity through policy actions and economic and non-economic incentives. Finally, the last section examines how biotechnology can contribute to the conservation of its most valuabie input and discusses some of the more important contributions to date.

Biological diversity or the total variability of life on earth was a term coined in the early 1980's. Concern about biological diversity and the destruction of habitats rose along with a greater awareness about the environment in general and its relationship with the human civilisation and other species.

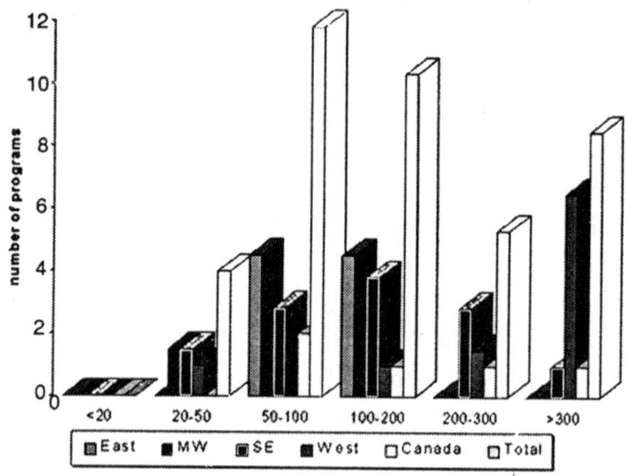

Fig. Ecological diversity

Another definition of biological diversity that has been widely used is that contained in Article 2 of the Convention on Biological Diversity:

'Biological diversity means the variability among living organisms from all sources, including *interalia*, terrestrial, marine and other aquatic ecosystems and the ecological complexes of which they are part; this includes diversity within species; between species and of ecosystems'.

Species therefore make up the basic unit of classification in biodiversity and diversity can be described at different levels such as genetic diversity and ecological diversity. *Genetic diversity* is diversity at the most micro level and refers to variability within and between populations of species. It can be measured in terms of differences between DNA sequences for example. *Ecological diversity* refers to the variation between species within a certain geographical area or ecosystem. Thus, while research on genetic diversity would tend to focus on the loss of genetic variability in crops and animals, research on ecological diversity would concern itself with changes in compositions of species within particular regions.

The Convention on Biological Diversity (CBD) came into force at the end of 1993. It has been signed and ratified by around 160 countries and reflects a desire to better understand the interaction between species and their environments and to try and reduce the rate of decline of biological diversity. The Convention's focus is not simply on the conservation and preservation of biological diversity.

Instead, its broader goals such as the sustainable use of the components of biological diversity and the equitable sharing of benefits acknowledge the importance of interaction between the users and preservers of biodiversity. Indeed, in many cases, the users have become the preservers of biodiversity. It is also the goal of the Convention to find ways by which the new users of biodiversity, namely the biotechnology industry and the older users of biodiversity such as indigenous populations, could help to conserve biotechnology together.

Importance of Biodiversity

Biodiversity is not only of critical importance for those communities that have become its direct users. Rather, changes in species and ecosystem diversity have the potential to bring about radical changes to the earth's environment and its peoples. The extinction of one species will often lead to the extinction of others. Because of the interdependence between species, the impact will be felt not only on those species that were directly dependent on the extinct species, but also on those that were indirectly dependent.

A slight change in the ecosystem may affect species diversity and it is more likely that the presence of many species will increase the chance of some species being able to tolerate the change. Monoculture plantations on the other hand may be more vulnerable to disease or climatic change. Biodiversity also holds a more tangible value for its users. Farmers and plant breeders have, for example, relied for centuries on local varieties found in the wild to develop and cultivate the domestic varieties of crops that are consumed today. Similarly, in pharmaceuticals, it is estimated that over 5,000 species form the basis of Chinese traditional medicine and 2,000 species have been used by Amazonian populations for medicinal purposes, while in modern medicine more than 3,000 antibiotics alone, including penicillin and tetracycline, are derived from micro-organisms.

The establishment of industrial biotechnology today as an exciting new technology, with an enormous potential for developing cures for hitherto incurable diseases and for expanding food production in an ever expanding world, has placed new demands on the world's genetic resources. Now, more than ever, there is a need to understand and classify the world's biological diversity so that it can be utilised in a sustainable manner.

The Value of Biodiversity

One of the most difficult problems raised in the quest to ensure the sustainable use and conservation of biological diversity is that of adjudging an appropriate value to natural resources. The 'heritage of mankind' as it is known, is in fact a public good, i.e. it is freely available, and therefore suffers from what economists describe as market failure, where market prices, based on private value, do not accurately reflect its true value to society.

While economic values for biological resources may exist, based on their immediate use, no value is placed on the depletion of biodiversity or on the knowledge that derives from it. Governments on their part have also failed to acknowledge the usefulness of this resource and to develop adequate incentive measures to prevent its depletion. Policies which encourage development of one sort may directly or indirectly have negative impacts on biodiversity and the functioning of ecosystems.

For example, price controls and/or subsidies in agriculture, water, energy, transport and communications sectors can have adverse impacts on biodiversity. Similarly, Governments often do not invest efficiently in land development, allowing marginal land to remain available for conversion, despite the inefficiency of such conversions. Such policy failures which have failed to close the gap between the value of biodiversity to private users and its value to society as a whole have, therefore, exacerbated the problems of depletion caused by market failure.

The world's biological base appears quite obviously to be a valuable resource. It has contributed to agriculture, medicine and supports a range of other activities. Moreover, its contribution does not appear to be finite. Scientists and farmers are constantly using germplasm to improve agricultural stocks or medical cures. For example, it has been claimed that of the 5,000 or so plant species that have been screened for their medicinal value, around 40 of these are presently in use in prescription drugs. The healing properties of many others are used regularly in indigenous medicine or in the home. In agriculture, similarly, plant screening to identify improvements to plant productivity and disease and pest resistance is done on a regular basis, both to develop new varieties, and also to maintain stocks of present varieties.

Another reason why continuous research and therefore continuous utilisation of biological material has to occur, is the evolving nature of the world and our environment. Plant breeding and medical researches are often required to be reactive to new diseases and climatic changes. New strains of diseases, for example, tuberculosis and malaria in recent years, ensure that research even on diseases with widely used and available cures will send scientists back to germplasm stocks to investigate their properties.

It is therefore not simply the market value of biological materials as and when they are used in productive activities that need to be ascertained. The information value contained in biodiversity, instead, is potentially more valuable to the scientist and society at large. Furthermore, we simply do not have adequate understanding of how economic activities affect ecosystems and what the consequences of biodiversity loss may be. Rejuvenated interest in the value of biological resources through the rapid advances made by biotechnology in recent decades should therefore, if anything, increase further the value of biodiversity and certainly heighten one's awareness of its importance, although this does not make it any easier to give biodiversity an appropriate economic value.

The Responses

Thus far we have attempted to show why conserving biodiversity is important and necessary. Various institutional and other mechanisms have been set up or can be set up for conservation. There are, however, two basic problems we face: first, to identify what we mean by biological diversity. As discussed in section 6.2, biodiversity is not a homogeneous resource. There are complex overlaps and interdependencies between species and ecosystems. Depending on what we mean by biodiversity, then, the set of solutions to the problem of conservation will be considerably different.

The second problem is one of defining an appropriate set of instruments that can be used to ensure sustainable use and conservation of these identified resources. One approach is to use instruments known as 'biodiversity indicators', for example indicators that show the level of and diversity of wild species in a particular habitat. While the approach is undoubtedly useful, it leaves much up to individual Governments and policymakers on which indicators to give priority to and therefore, by default, to define which aspects of biodiversity are particularly desirable and must be conserved. On a more 'social' level, it would perhaps be most desirable to define criteria for measuring and valuing biodiversity that are somehow global in scope.

Since as discussed above, the erosion of biodiversity is caused by a combination of market and government failure, it has been suggested that the most appropriate means by which to encourage the conservation of biodiversity is to use both policy measures and economic incentives to try and increase awareness about the destruction of biodiversity and encourage its conservation. Article 11 of the Convention on Biological Diversity calls on contracting parties to 'adopt economically and socially sound measures that act as incentives for the conservation and sustainable use of components of biological diversity'. There are now numerous studies on incentives for biodiversity conservation.

Thus, along with creating economic incentives to increase awareness and encourage biodiversity conservation, governments have also established national parks authorities, Government ministries for the environment and other institutional mechanisms to focus on the problem of biodiversity depletion. However, given the global nature of the problem and the desirability of having all nations taking part in the conservation effort, international agreements to implement and finance conservation efforts have also increasingly been set up.

Individual conventions or agreements have been set up focusing on a particular region or a particular problem, for example CITES, which restricts trade in endangered species as a way of reducing the rate of extinction. The global effort to reduce the extinction of biodiversity however, culminated in the all-encompassing Biodiversity

Convention, which was opened for signature in 1992. This, along with other efforts to reduce the rapid rate of depletion of biological resources, is the subject of the following chapter.

Table: Some examples of common incentives

Regulations	These include harvesting permits, collection permits, land clearing permits, the creation of development zones and precautionary standards and regulations
Economic price based incentives	Negative incentives such as fines and taxes can be accompanied by positive incentives such as awards, grants and subsidies. In addition, bioprospecting activities can be used productively to identify and add value to biological resources.
Institutional mechanisms	These play an important part in ensuring that regulations and economic incentives are implemented and are complementary to each other. They include non-governmental organizations, national departments and ministeries for the environment, national parks and on a global level, multilateral environmental agreements.

RELATIONSHIP BETWEEN BIOTECHNOLOGY AND BIODIVERSITY

Biotechnology and biodiversity are related to each other most directly through use: biotechnology benefits from and makes use of the world's biological diversity. However, the relationship is not unidirectional and it is an understatement to define it as complex. While the goal of this chapter and the following is to examine the manner by which biotechnology can make a positive contribution to biodiversity conservation, it is firstly important to recognize that until now the relationship has been rather conflicting.

The use of biotechnology and other biological technologies before it to introduce new plant varieties in agriculture has resulted in a controversial debate. For those who pioneered and participated in the development of green revolution technologies in the 1960s and 1970s, the benefits of hybridization and high yielding varieties were obvious: higher productivity, and improved disease and pest resistance, allowing countries to become selfsufficient in grain production and able to feed their poor populations.

However, for others, one of the consequences of the introduction of high yielding varieties has been the loss of biological diversity. The introduction of new high-response varieties led to the displacement of many traditional varieties. In addition, it was argued that the introduction of high-response varieties of wheat and rice displaced traditional crop variety, resulting in the creation of vast areas of monoculture.

The literature on this subject ranges from studies carried out on green revolution technologies, to the early impact of biotechnology in agriculture. It has been estimated that local communities maintain a much higher degree of genetic variety in the crops they cultivate, than commercial agriculture, which relies on a much narrower genetic base, utilizing only a few modern varieties, thereby increasing the susceptibility of modern agriculture to attacks from pests and diseases.

The isolation of characteristics that are beneficial, such as disease and pest resistance, response to climatic variations and higher productivity characteristics, although ultimately beneficial for international consumer welfare, is the further cause of reduced genetic diversity according to this argument. Indeed, there are examples in the literature of entire species being lost as a result of monoculture in planting as well as the uniform characteristics of the plants replacing them.

The second source of controversy relates to the value of biological diversity, especially in relation to the value of biotechnology based products that make use of biological materials. Biological material has traditionally been treated as the common heritage of mankind and therefore accessible by all. Its valuable properties have contributed not only to agriculture, but also to medicine and a number of industrial processes. Conservationists, however, argue that despite its immense contribution to the progress of human civilization, the value of biodiversity is not sufficiently acknowledged and is therefore the proximate cause of its continued deterioriation.

Here, the conflict between developing and industrialized countries is at its most acute. As the number of patents granted for biotechnology products and processes increases, developing countries are demanding that the value of indigenous knowledge and the properties of biological materials that were utilized in the production of the biotechnology invention be acknowledged. The opposition expressed by conservation groups to recent patents granted for products based on the healing properties of neem and turmeric, has highlighted some of these differences.

Developing countries have also argued that while they have been *de facto* made responsible for the preservation of the world's biodiversity, most of which finds itself in the developing world, they are unable to benefit from the use of their own biodiversity because its contribution and that of local knowledge is not acknowledged by those who develop new biotechnology products, most of whom are based in the industrialized world.

Many have threatened that unless they receive some form of compensation for the use of their natural resources and of local knowledge, they will not allow access to those natural resources by biotechnologists or plant and animal breeders. Some countries have gone further, and have tried to halt the outflow of germplasm by making the export of some of these products illegal. However, the cost of monitoring their outflow has been much higher than the immediate benefits and many countries are unable to enforce their regulations.

Thus, the conflict between the biotechnologists and conservationists appears in the first instance to be unresolvable because of fundamentally different approaches to the question of property rights and economic and social values relating to biotechnology and biodiversity. Nevertheless, the validity of both arguments makes cooperation rather than conflict between the two sides inevitable — inevitable because the survival of both biotechnology and biodiversity in the long run is dependent on the support of the other. The next section suggests a number of contributions that can be made by biotechnology in the conservation effort, before the discussion moves on to the practical contribution of biotechnology to this effort.

The Contribution of Biotechnology to Conservation Efforts

While much of the literature on the relationship between biotechnology and biodiversity has examined its conflictual nature, the relationship can play a positive role as well.

First, biotechnology provides the scientist with powerful new tools to identify and assess biological resources. This is used increasingly in activities to identify new compounds which can be utilised in the biotechnology industry.

Second, those very same tools that are used to transform biological resources into biotechnology products, can be used to conserve biological diversity.

Third, biotechnology also provides a methodology for using biodiversity efficiently as a genetic resource for production and also for remediation of ecosystems which have been destroyed.

Identification and Assessment of Biodiversity

At its simplest level, the search for 'commercially useful' biological materials by the biotechnology industry produces a record of these materials and also of those characteristics that are useful for the biotechnology industry. For example, a number of patents that have been registered for biological inventions and are based on such characteristics found in biological materials, are often registered along with the patent and those characteristics relevant for the invention are described. For example a patent registered with the European Patent Office contained the following description:

'Patent no. EP 0513671: The present invention relates to new therapeutic applications of extracts, fractions and single active ingredients prepared from *Cammiphora mukul*: the invention further relates to processes for the preparation of the total steroidal fraction which is present in the exudate of the above plant. The *Commiphora mukul* (Hook ex Stocks) Eng. (syn. *Balsamodendron mukul Hook*) is a small tree of the Berseraceae family, endemic in the Indian peninsula, growing spontaneously in the dry and semi-dry Rajasthan, Gujarat and Madhya Pradesh districts in India and in Beluchistan district in Pakistan'.

Similarly, in the case of patents filed at the United States Patent Office, reference to origin can also be found:

'Patent US 3743722 which refers to the anti-coagulant properties of two pit vipers: these species are found in different parts of the world, predominantly in Southeast Asia (*Agkistrodon rhodostoma*) and South America (*Bothrops atrox*)'.

Local use Value and Knowledge

Implicitly, patent disclosures, therefore, exhibit their reliance on previous 'public domain' knowledge based on common usage. Ethnobotany, based on indigenous or local knowledge is also used increasingly by biotechnology companies. The Convention on Biological Diversity explicitly recognises the importance of such local knowledge based on centuries of local use, and calls signatories to 'respect, preserve and maintain knowledge, innovations and practices of indigenous and local communities ... and promote their wider application with the approval and involvement of the holders of such knowledge...'.

The present debate revolves around devising methods by which to more explicitly acknowledge the contribution of this knowledge to modern innovations and also to conservation and sustainable development. The problem, however, is one of reaching a consensus on what value to give this knowledge which lies essentially in the public domain and is therefore 'free to be utilised'.

Assessment Methods

As biotechnology moves to higher levels of sophistication, it allows scientists to investigate the diversity of organisms and their ecosystems through a detailed examination of the genetic makeup of organisms. Such an examination is the first step in identifying and selecting for desirable characteristics in that organism. In their search for appropriate and beneficial characteristics, scientists sample and screen thousands of plants on a regular basis, 'thus identifying key genetic characteristics in plants and other organisms. Biological collections, either in situ or ex situ are therefore likely to be maintained in the long run because they provide such an abundant and useful resource for the biotechnology industry.

Along with increasing sophistication in biotechnology, the key methods of obtaining information on biological materials have also become more advanced. Some of those presently in use include chemical screening which classifies organisms in terms of their chemical phenotypes. The recent development of high performance liquid chromatography has also simplified the process of separating molecules on the basis of their physical and chemical properties; and the use of DNA based technologies provides details of the genotype of the organism.

This is what the genome mapping effort is trying to achieve with regard to the human being. Examples of the use of such technologies include DNA fingerprinting.

Biotechnology can also make a contribution to the assessment and management of biological diversity through taxonomy and a systematic collection of information relating to the relationships between organisms and the impact on them of changes in their habitats.

Using Biotechnology to Conserve Biodiversity

Apart from identification and taxonomic activities, the tools of modern biotechnology can also be utilised to provide better methods for conserving biological diversity, both for the immediate use of the biotechnology industry and also for the long-term sustainability of ecosystems.

In Situ Conservation

As the CBD points out, the optimal environment in which biological resources can be conserved is *in situ*, i.e. within the ecosystem that they originally come from. There are several reasons for this method being optimal. First, if a species is conserved outside its habitat, only certain aspects of it can be conserved and not all of its diversity. Second, if only one or a few species are conserved out of an entire habitat, we will never be certain how much diversity and how many valuable species were lost as a result of the destruction of that habitat. Third, we are simply not aware of the extent of interaction and interdependence between species within their natural habitats and whether species diversity can be maintained outside of that habitat, once deprived of that complex interaction.

Until recently, *in situ* conservation was seen as the problem of national and local governments and their conservation programmes. Conservation activities have often been carried out by marking out certain areas of high biological diversity and laying down strict guidelines on use to ensure their conservation.

However, with the greater need for screening diverse species of plants and animals, the biotechnology industry has realised the need for investing in such conservation efforts. A number of private companies and individuals have started to invest directly in such conservation projects to ensure the survival of this diversity.

Ex Situ Conservation

Although *in situ* conservation is regarded as an optimal method of conserving biological diversity by many, it is often not feasible for a number of reasons such as habitat destruction and changes in the ecosystem which make species survival unviable. For this reason, *ex situ* conservation for example in gene banks, zoos and botanic gardens has become the second best method of conserving biodiversity.

Simpler biotechnologies such as tissue culture have been used to maintain the quality and viability of these collections. Despite this, the viability of this 'banked' material is being questioned. In 1991, representatives of 13 national germplasm banks in Latin America reported losses of between 5 and 100 per cent in batches of maize seed collected between 1940 and 1980 making them unviable.

Modern biotechnology techniques can be used to provide a more systematic characterisation of samples on the basis of their DNA rather than through the storage of seed which may or may not germinate at a later date. DNA libraries are now being built by a number of companies to provide an analysis of the genetic structure of materials collected and also to be able to relate them to similar phenotype traits in other organisms. While these are not substitutes for collections of *in situ* and live collections such as in botanical gardens and zoos, these modern methods complement those of traditional conservation programmes and will provide necessary information on potentially useful traits several years from now.

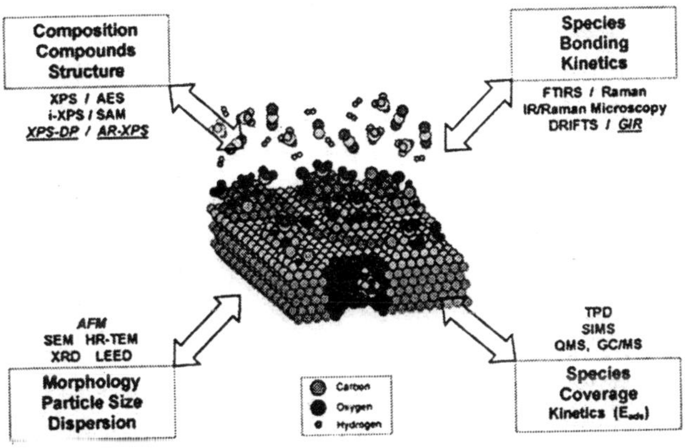

Fig. In Situ and Ex situ

Sustainable Use

Increasing environmental awareness has made us increase our use of biological materials in economic activities reducing the strain on the environment. For example biological material is being used to leach ores from mine sites. In many cases these ores are just dumped at the mine sites because extraction may no longer be economical through the use of conventional mining techniques. Improved management of a natural biological process has however demonstrated enormous potential. Research in this area has also led to the use of biological materials in remediation of industrial and metallic wastes thereby lowering pressure on the environment and reducing the cost of cleaning up polluting effluents.

The depletion of biological diversity does not and must not, however, mean a halt in its use. The contribution of this rich and diverse resource to modern medicine and agriculture has been immeasurable and must continue to contribute to future generations. The use of biological diversity by the biotechnology industry, moreover, is unlikely to end because of the accusations levelled against it by conservationists.

Instead, biotechnologists must increasingly be convinced of the importance not only of using the resource to develop miracle cures, but also of using the resource in a sustainable manner so as to be useful for future generations. The biotechnology industry also, arguably, has the resources and the technology to contribute to sustainable use and conservation activities as discussed briefly above.

Indeed, in many cases, the users of biodiversity are indeed making significant contributions in this direction. The scientific development of biotechnology, followed by its technological commercialization has followed very different patterns in different countries. The intention of the book was to compare and contrast the development of biotechnology in different countries with different levels of scientific and technological capability. In doing so, we first examined the development of the technology among a group of industrialised countries with similar levels of scientific and technological capabilities, followed by the experiences of a group of technological followers.

The experiences of all the countries are dependent on their scientific strengths, their economic and social environments and their technological and industrial structures. However, there has been a tendency among countries, be they industrialised or developing, to follow patterns of development established in other countries where biotechnology was introduced earlier.

Thus, the spread of biotechnology in industrialised countries such as the United States, in the 1970s and 1980s, led to the establishment of 'biotechnology programmes', or 'biotechnology directorates', in many countries to hasten the diffusion of the technology. Many industrialised and developing countries established nationwide programmes for the development of biotechnology during the 1980s, which aimed to define national agendas and priorities for biotechnology.

This was followed by attempts to improve the commercialisation of biotechnology, following patterns established primarily in the United States. Thus, the small biotechnology company became the key to commercialisation and venture capital the key to finance. Most European countries have active programmes to encourage the development of small biotechnology companies which are often believed to have been more successful than other incumbent firms in this field.

Similarly, a number of developing countries have also established programmes to encourage venture capital financing and small biotechnology companies. However, what has been the impact of these programmes and replicated measures for biotechnology development and diffusion? Have the institutional mechanisms thus created helped industrialised and developing countries to enhance their scientific and technological capabilities?

Development of Technological Capabilities

Comparing the development of biotechnology in developing and industrialised countries, therefore, one can identify a number of rather simplifying patterns:

First, the absolute levels of scientific and technological capabilities are much lower in the developing countries. Scientific education and training programmes, moreover, tend to restrict movement and interaction between the sciences much more than in the industrialised countries, making the diffusion of an inter-disciplinary science such as biotechnology difficult through the educational system. For governments the obvious problem in changing the educational system to accept new techniques such as biotechnology, as well as introducing biotechnology courses within regular curricula, is a lack of resources to expend on such programmes.

Second, and perhaps much more problematic, is the level of commercial applicability. While in the industrialised countries, commercial application of biotechnology has been faster in some than in others, commercial applicability in the developing countries has been hindered by a large number of factors. In industrialised countries such as the United States, where barriers between the scientific and industrial community are less well defined than in several other countries with similar levels of scientific capabilities, we saw that the small biotechnology company led the development of biotechnology.

Other factors that were helpful in the early development of commercial biotechnology included the availability of funds such as venture capital and later application for public stock through the Nasdaq. In Europe and Japan, where such funds have been less readily available and the barriers between the scientific and industrial communities higher, commercialisation has been a more difficult task. However, there is ample evidence to show that new patterns of commercialisation are emerging, often to replace the gap left by the (lack of) the small biotechnology company.

For example, in Europe, large companies, notably well established multinational corporations are utilizing their resources and rising to the forefront of biotechnology research; and new mechanisms are being developed for technology sharing and transfer, such as the formation of technology or marketing alliances with other companies or scientific institutions.

In contrast, in most developing countries, these funds and similar mechanisms simply do not exist, making commercialisation of risky new technologies extremely difficult. Again, the Government has had to play a role in developing these sources of funding for entrepreneurs, often unsuccessfully.

Third, the problems that biotechnologists are attempting to tackle in developing countries are relatively different from those in the industrialised countries making the acquisition of scientific knowledge and technology essential. While these problems can, to a certain extent, be addressed through technology transfer, in many cases they have to be adapted to local conditions and needs. In addition, the needs of developing countries to document and conserve their biological diversity makes the acquisition of knowledge, technology and the development of local technology and human resources important.

The scenario that faces policymakers, scientists and industrialists in developing countries, therefore, is different from that of their counterparts in industrialised countries. Their solutions to their challenges have accordingly been very different. The relative lack of commercial breakthroughs in developing countries has made the role of the Government more prominent in giving policy and research direction to biotechnology.

One can examine the different problems and solutions chosen by industrialised and developing countries at two different levels — the external regulatory framework within which companies and scientists must function and the differences in overall institutional and industrial structures within which biotechnology is commercialised.

The External Environment

The external regulatory and enabling environment in which scientists and industrialists work becomes all the more important as the needs of new technologies render older structures obsolete. In the case of biotechnology, apart from the general economic environment which helps in the diffusion of the technology, the presence of guidelines establishing the rights of researchers and the general public become defining factors in its development.

The establishment of regulatory guidelines, including legislation on intellectual property rights, which in many cases is non existent, and the regulation of use and release of genetically engineered material, has become an important aspect of the Government's role in the development and diffusion of biotechnology.

The usual rationale for Government intervention in the economy is in the case of a public good, where the market will fail to allocate resources effectively for the benefit of society. However, very often the object of the intervention is not necessarily a public good and market failure is exacerbated and replaced by Government failure. In the case of biotechnology, once the private sector became actively involved in commercial production, the role of the Government in the commercial sphere became less obvious.

Government policies were initially used to define national programmes and priorities for biotechnology. As scientific capabilities were transformed into biotechnology based products, the focus of biotechnology policies shifted from developing basic capabilities in science and technology to the problems of commercialisation. Many governments began establishing mechanisms to facilitate technology transfer from basic research institutions to firms.

Governments have even established biotechnology firms to act as conduits for technology developed in research institutions or Government laboratories. Some governments have focused on the role played by certain types of firms, notably the small scale dedicated biotechnology firm which was highly successful, and remains so in the United States, and venture capital and other financial funds have been established to hasten the development of such firms dedicated to high risk technologies.

Regulation

More essential, however, has been the role played by the Government in developing regulatory structures for biotechnology research and development. As with other technologies before it, the development of biotechnology products and processes has identified gaps and failures in the present regulatory system developed by countries. As with other technologies before it, maximum access has to be ensured to the scientific community, while at the same time protecting the rights and security of society in general.

Biotechnology, especially due to its ability to manipulate genetic material, has raised a number of sensitivities among scientists, conservationists and the general public. It raises a special need for regulatory issues to address all the concerns of these groups and achieve the right balance which ensures technological progress.

Regulatory problems raised by biotechnology are not just seen in developing countries, but also in the industrialised world. For example, the European Union has had great difficulty in reconciling the different levels of regulation provided by different member states in one single piece of legislation. In developing countries too, the problems associated with public opposition to genetically manipulated material and the ethical and economic issues raised by such debates will become more apparent in the years to come.

Intellectual Property Rights

Similarly, as with all new technologies, the right balance is required between the protection of inventions versus protection of society's right to use and have access to protected material. This may be different in different societies. Ignoring the issue of patents or other *sui generis* forms of protection may only be becoming counterproductive to a country's objectives. The recent instance of a patent being granted to a company based in the United States for a variety of Basmati rice, is a case in point. The Indian Government has declared its intentions to contest the patent on the grounds that Basmati rice is grown in a region in India and Pakistan and hence entitled to protection under legislation relating to geographical indications.

Hence, Basmati rice, as with Swiss cheese or champagne must come from a certain geographical region of the world to be called as such. However, while other products have been accorded such protection, India has no legislation for geographical indications and has thus far resisted introducing such legislation. Basmati producers may thus find their claim of being the only ones entitled to call their rice Basmati defeated because of the failure of their Government to introduce protective legislation.

Countries such as India, which use biotechnology for farming as well as industrial and pharmaceutical processes, may increasingly be in a position to require some form of intellectual property protection to encourage innovation rather than only imitation. Many developing countries such as India also have substantial bodies of indigenous knowledge based on traditional uses of biological materials which they would like to protect.

The protection of indigenous knowledge and the rights of indigenous peoples is provided for by the Convention on Biological Diversity (Article 8(j)); countries can also design their own *sui generis* systems for protecting biotechnological inventions under the WTO's trade related intellectual property rights (TRIPS) agreement. This would also enable them to address their needs to develop biotechnology and to accelerate activities to conserve biodiversity.

Education and Training

Another aspect of government policy which continues to be vital for biotechnology development and biodiversity conservation, is training. In developing countries especially, a lack of sufficiently trained human resources has often been identified as a serious barrier to technology development. Moreover, it is also a barrier to developing techniques by which to conserve biodiversity. Training and education retain public good features which must be addressed through government policies and resources.

At the most, these policies and resources would only be supplemented by the private sector, and often only once a basic infrastructure is already in place. Thus the role of Government policy and intervention in the development of biotechnology has been evolutionary in most industrialised countries, moving from a more interventionist role in setting up venture capital based funds, or companies, to a regulatory and institution building role. In fact, in many European countries, where the Government has tried to play an active role in closing the gap between basic and applied research or commercialisation, the evidence does not necessarily show favourable results.

In developing countries, while the role of the Government will continue to be more prominent due to a relative lack of supportive structures for biotechnology development, there is also increasing evidence that its actions in the field of regulation are becoming more important, both for protecting the health and rights of its citizens and also for furthering the commercialisation of biotechnology.

Contrasting Industrial Structures

The acquisition of scientific research capabilities through indigenous research or through technology transfer and reverse engineering, is only the first step in the complex process of entry into the market for biotechnology. Market entry with the right product or technology, at the right time, has proven difficult for companies in both industrialised countries and especially in developing countries.

However, while the evolution of Government policy has broadly followed similar patterns in industrialised and developing countries, an appropriate pattern of development of industrial structures is less obvious. The experiences of industrialised and developing countries have been relatively different thus far.

Experiences in Industrialised Nations

It has been noted in this book that amongst the industrialised countries, the best contrasts are offered by the United States and the European Union where, with similar levels of scientific capabilities, the commercialisation of biotechnology has followed a different pattern and pace. The relative success of the small biotechnology firm in the United

States and later in parts of Europe, notably the United Kingdom, has led European Governments as well as the European Union to set up special assistance programmes for small companies in the field.

One noted that the types of structures used to commercialise biotechnology varied widely between the small biotechnology company that has been more successful in the United States, and more recently in the United Kingdom, and the multinational firm whose strong grounding in scientific and industrial research in many European companies has made it more successful in some cases. It is, moreover, unclear that the commercialisation of biotechnology by one or the other institutional structure is more or less successful. Much depends on the internal strengths of companies, as well as that of the external environment.

Thus, the obvious strength of the large multinational firm in Europe has been its background in scientific research and development but also the lack of a flexible financial environment which has proved disadvantageous to the development of small biotechnology companies.

Experiences of Technology Followers

In the case of less technologically advanced countries, it is even less clear that they should be following the example set by the United States or other countries. Most developing countries, apart from having a relatively weaker scientific base, lack the most basic infrastructure for commercialising risky new technologies.

The gap between academic research and the industrial community is even more pronounced; financial capital sources are practically non-existent, especially for risky new technologies such as those associated with biotechnology; and the market for many of the risky new technologies is not well defined.

Given these factors, it is considerably harder for new firms to break into these markets. Incumbent firms, moreover, are not globally competitive in research and development. Their choices of technology and technology alliances are therefore restricted to problems that are local in nature. Their solutions to the problems must therefore be dependent on a good understanding of the dynamics of biotechnology in their own particular environments.

For example, we saw that the relative risk averseness of the private sector, combined with a lack of financial capital for high risk new technologies has resulted in an incremental approach to the commercialisation of biotechnology. The private sector has been most successful in incremental technologies such as tissue culture, in agriculture, and other second generatión biotechnologies, in contrast to government policies that have encouraged the development of genetically engineered products which requires a larger technological jump.

Many of the technologies favoured by the private sector in developing countries are based closely on more recent technologies, such as those developed during the green revolution of the 1970s and older techniques, building upon knowledge already present in these countries.

The development of incremental technologies based upon previous knowledge is also likely to increase interaction between biotechnologists and the indigenous community, helping both communities to understand the interlinkages between the development of biotechnology and the conservation of biological resources.

Thus, the transition from basic research to commercialisation is a process that is dependent not only on the ability to achieve a certain level of technological capability, but also on the conditions of market entry and existing industrial structures and institutional history. All these factors will vary from country to country and will change over time as institutional relationships change. Governments would do well, therefore, to provide a more neutral incentive and support structure to all participants in biotechnology research and development rather than trying to build new institutional relationships which may not be suited to biotechnology development in that particular environment.

Chapter 10

Biotechnology in the New Millennium

TECHNOLOGICAL CHANGE, INSTITUTIONAL CHANGE, AND POLITICAL STRUGGLE

In September 2000, the business section of the *New York Times* ran a frontpage story about a new problem in the nation's food supply. A new genetically engineered (GE) corn containing the protein CrygC had been found in several nationally sold brands of corn tacos. Although the corn, known as StarLink, had been approved for use in animal feed and in industry, it had not been approved for human consumption. Yet when a group of social activists hired an Iowa-based company to test the tacos, there it was.

What seemed at first to be just another bump in the history of biotechnology turned out to have enormous repercussions. In addition to ending up in the U. S. food supply, where it wreaked havoc, the genetically engineered corn found its way into the food systems of many other countries in which it had also not been approved. Tensions over trade in GE agricultural products flared, and key U. S. agricultural importers, such as Japan and South Korea, began rejecting shipments of U. S. corn on the grounds that they were contaminated with StarLink. When U. S. agricultural importers began to seek non-GE sources of corn from such countries as Brazil and China, U. S.

farmers and the U. S. Department of Agriculture (USDA) panicked about the major market losses that would inevitably result.

In the U. S. heartland, where most of the corn had been grown, the contamination problem exploded in another way. The company that produced the seed, Aventis CropScience, tried to blame farmers for not following proper planting procedures and not taking adequate care to keep StarLink corn separate from their conventionally grown crops. Angered and offended, farmers charged that the company had not properly informed them about how to raise and segregate the corn. Both groups were deeply concerned about the liability for losses associated with StarLink. The incident created a rift between these formerly friendly business partners and revealed differences in their interests in and positions on the technology, differences that had not been previously apparent.

Some five months after the unapproved corn was first detected, the USDA agreed to spend up to $20 million to buy back the remaining StarLink corn seed from seed companies. The Bush administration's rationale for the buyback was that the debacle was in part the fault of the U. S. Environmental Protection Agency (EPA), which had agreed to a two-tiered system of regulation on the assumption that GE corn could be grown for animals and kept out of the human food supply. Despite the administration's efforts to resolve the crisis, reverberations from the StarLink incident were felt long after the problem was identified. In the end, food manufacturers recalled nearly three hundred products, U. S. corn trade with many foreign countries was seriously disrupted, and the industry paid more than $10 million to farmers in Iowa alone.

In March 2002, a federal judge settled a class action lawsuit, brought by consumers who complained of allergic reactions to StarLink corn, with an award of $9 million. But perhaps most significantly, the StarLink incident moved agricultural biotechnology squarely into public view. In many ways, the StarLink corn incident is emblematic of the political, economic, and institutional crises that have befallen agricultural biotechnology at the beginning of the twenty-first century. As we move into the new millennium, agricultural biotechnology—that is, the use of recombinant DNA techniques in food, feed, and raw materials—faces a highly uncertain future.

Although there is little doubt that biotechnology as an industrial production complex, a set of material techniques and practices, a cultural icon, and a bundle of social relations is here to stay, the directions in which the technology is likely to go, the extent to which it will become part of our daily lives, the uses to which it will be put, and the political and institutional environment in which it will evolve are more uncertain now than at any time since the early 1980s. Indeed, we appear to be standing at a crucial juncture in the history of the technology, a juncture at which the future is surprisingly open to the actions of a broad range of (human and nonhuman) actors.

Although the StarLink incident can be seen as the high-water mark in the contemporary crisis of agricultural biotechnology, the causes of the crisis are complex and involve deep tensions that have been brewing for decades. These tensions reflect a clash of worldviews about whether such a potentially revolutionary technology should be introduced into the socionatural world at all; who should have the power to decide; what kinds of precautions should be taken to limit the harm that is done to people and the environment; and what kinds of institutional structures should be established to control and regulate these technologies. As a geared-up and heavily invested (in all senses of the word) life sciences industry has moved forward with its technoscientific interventions in virtually every form of life, many have started to challenge the assumption that any one segment of society should have such a prerogative. These challenges have made genetic engineering into one of the most contested technologies of our time.

While social resistance is one source of uncertainty for the biotechnology industry, farmers, food retailers, and policymakers represent another. Will farmers continue to find a market for their products if they use these increasingly controversial technologies on their farms and in their dairies and fish pens? If farmers opt not to use the technology, will they be adversely affected by genetic "contamination" from pollen drift or other sources of genetic change or by the presence of trace amounts of genetically engineered organisms (GEOs) in the food supply chain?

Will retailers lose customers by allying themselves with the biotechnology industry instead of with some of their most vocal and concerned consumers? Even government regulators are unsure how to deal with the political sensitivity of these new technologies. Should regulatory agencies continue to facilitate their commercialization, or should they respond to that portion of the public that advocates that we proceed with greater caution? Different groups are clearly worried about different issues, but all contribute to a growing uncertainty about biotechnology at the beginning of the millennium.

One of this book's primary goals is to illuminate the dynamics of this key historical moment, including the concrete political-economic changes taking place in and around the biotechnology sector, the institutional foundations on which those changes have been built, and the political struggles they are generating. Ten to fifteen years ago, a spate of studies predicted a number of important social, economic, political, and environmental changes that would occur as the science and industry of biotechnology developed. These changes included the emergence of a new and closer relationship between industry and universities; the erosion of biodiversity; growing social and economic inequalities associated with the patenting and use of these new technologies; and their potentially adverse environmental consequences for ecosystems.

Although the industry was still young at the time of these studies, they pointed to the way in which biotechnology was leading to greater economic concentration in the agro-food sector to the privatization and commodification of plant genetic resources and other life forms; and to the further extension of capitalist production relations into the farm sector. The best of these studies took a historical approach to these issues but also identified what was new about the new biotechnologies.

Although this body of literature proved prescient in identifying some of the critical changes biotechnology would introduce into society, the biophysical world, and the global agro-food system, three important new developments call for a fresh look and additional analysis. First, the technology has moved from the lab to the field—or more accurately, from the lab to the field, the barn, the forest plantation, the sea, and the human body. Between 1996 and 2001, the number of acres planted with GE crops— almost exclusively corn, soy, canola, and cotton—rose from an estimated 1.7 million to an estimated 130 million worldwide. Genetic engineering techniques are also being applied to a growing number of life forms, including insects, farm animals, marine organisms, trees, and humans. Perhaps more important, technology development has been revolutionized—and greatly accelerated—by the advent of genomics and the synergies that have emerged between molecular biology, recombinant DNA techniques, and the bioinformatics sector.

The second major development has been the extraordinary politicization of the technology. Just when the science and industry of biotechnology began to take off, the technology became a lightning rod for local, national, and transnational conflicts over trade, agriculture, and the environment. These conflicts and the political struggles they have engendered are transforming public perceptions (perhaps most importantly, among investors and consumers) of biotechnology's significance, promise, and future and have shaken the industry to its core.

The evolution of biotechnology, and the public's reaction to it, have strained existing social, economic, legal, and political institutions and revealed critical inadequacies in their ability to cope with the wide array of challenges posed by the technology. These processes are creating powerful demands for institutional change by specific parties and sets of interests. Thus the third major development has been institutional: as such weaknesses and problems are revealed, new institutional arrangements within and among societies—ranging from new food-testing systems to new international regulatory regimes governing trade in GEOs—are being created. These arrangements are changing norms and relations within the sector and are likely to have ramifications that extend far beyond it.

These developments raise many interesting questions, a number of which we seek to explore in this book. How has the biotechnology revolution actually affected the way we produce our food and raw materials? Has it altered the structure and organization of the agro-food sector and the power relations within it? What new social relationships have been forged as a result of the technology? There are also important questions about the institutional changes provoked by genetic engineering. What concrete institutional reforms have been associated with the technology? What are the dynamics of institutional change, and what political struggles are those dynamics engendering? What regulatory structures are becoming dominant within the emerging institutional matrix, and what groups and interests will they favour and empower? How will these new institutions shape the way we relate to the technology in the future?

We also need to understand better the politicization of the technology. What does this politicization mean, what are its consequences, and what is behind it? How broad-based is social resistance to biotechnology? Is this resistance confined to a handful of activist groups and organizations, or are we facing a rapidly expanding popular movement against the technology? How is the politicization of biotechnology influencing the trajectory of technological change? Although the number of academic and popular books being published on biotechnology has burgeoned, none has seriously explored these questions. It is this gap that the present collection seeks to fill.

TECHNOLOGICAL CHANGE

When the first wave of political-economic analyses of agricultural biotechnology appeared in the 1980s, the commercial application of modern biotechnology was still in its infancy. The technology clearly held the potential to transform the agriculture, food, human and animal health, and pharmaceutical sectors, but little technology had actually moved from university and industry laboratories into the commercial sphere. During the 1990s, however, that situation changed dramatically, as a large number of agricultural biotechnology applications were approved for use in the United States as well as abroad.

The two most extensively deployed biotechnologies have been those that render plants resistant to pests and those that confer herbicide tolerance to crops. Indeed, between 1996 and 1999, herbicide-tolerant soybeans were adopted more rapidly than any agricultural technology in the world, with pest-resistant corn not far behind. Reflecting U. S. farmers' embrace of the technology, in the year 2001, an estimated 68 percent of the U. S. soybean crop, 25 percent of the U. S. corn crop, and 69 percent of the cotton crop were planted with genetically engineered varieties. Also successfully commercialized in the 1990s were chymosin, a genetically engineered enzyme used in cheese production, and bovine somatotropin, a genetically engineered growth hormone that causes cows to produce more milk. Because GEOs are widely used by the food manufacturing industry, they have become virtually ubiquitous in processed food.

Although the adoption of GE crops has been extraordinarily rapid in the U. S., they have not spread around the world as fast or as widely as some of their proponents suggest. Indeed, although farmers in about a dozen countries legally planted GE corn, soy, canola, and cotton (the four most important transgenic crops) in 2001, the use of genetically engineered seeds remains heavily concentrated in only three countries: the United States, Canada, and Argentina. Together these countries account for more than 95 percent of the transgenic acreage planted worldwide.

Although it is mainly plant biotechnologies that have become a commercial reality in the last decade, molecular biologists and the biotechnology industry have applied their energies and resources to a broad spectrum of living organisms. Currently, scientists are genetically engineering animals for better disease resistance, faster growth, more attractive market traits (e.g., leaner meat or more meat), and higher-protein milk. Some animals—referred to as "pharm animals"—are being genetically engineered so that they can be used to produce body parts for humans and various human therapeutics (e.g., drugs or proteins) in their milk and tissues. Industry researchers are genetically modifying salmon to make them grow faster, and trees to improve market qualities. Even insects have not escaped the molecular biologist's gaze: silk moth caterpillars are being genetically engineered to produce stronger silk, male bollworm moths to render sterile the bollworm moth population in the southern United States, and mosquitoes to deliver vaccines to humans and livestock. Although none of these applications has been approved by the relevant government authorities and made it to the marketplace, much of this applied research is quite advanced, and some firms are now seeking regulatory approval for their organisms (e.g., transgenic salmon).

The Growth of the Biotechnology Industry

With the commercial deployment of the technology have come important changes in the size, significance, and structure of the industry. Economically insignificant twenty years ago, the biotechnology industry has become a leading sector in the economy. In the United States alone, the industry generated $20 billion in revenues in 1999, of which agriculture accounted for $2.3 billion. And these data include only U. S. companies that are *primarily* engaged in biotechnology activities. The inclusion of industries that provide inputs to the biotechnology sector or utilize the new techniques to produce their products would undoubtedly raise this figure much higher.

Martin Kenney describes the industrial complex that has grown up around biotechnology as a "new economic space."

This space has come to be populated by a handful of life sciences firms, most of which were formerly part of the chemical industry. Although early analysts clearly perceived the trend toward concentration, William Boyd (this volume) documents the breadth of the structural changes that have occurred as the industry, the technology, and the legal system have coevolved. In recent years, the life sciences firms have integrated vertically and horizontally and now encompass much of the commodity supply chain, from chemical inputs to seed companies to farming (in the form of production contracts). A study by Nicholas Kalaitzandonakes and Bruce Bjornson underscores the degree of concentration that has occurred: during the first half of the 1990s, there were some eight hundred mergers, acquisitions, and other strategic alliances in the agricultural input industry. There were only about a fifth as many a decade earlier.

Underlying this dramatic shift in industry organization has been the emergence of stronger intellectual-property protection for life forms. In the 1980s and 1990s, patent protection was broadened to include whole plants as well as their constituent parts (genes, gene fragments, proteins, and seeds). Patent rights were also extended, under certain circumstances, to DNA sequences. As the courts extended property rights deeper and deeper into the biological sphere, they created a powerful incentive for industry expansion.

There have also been changes in the nature of competition in the industry. These changes appear to have been driven by several factors—changes in property law, the fact that a handful of firms control the major segment of the industry, and the emergence of the fields of genomics, bioinformatics, and, most recently, proteomics. According to William Boyd's analysis, the gigantic firms that dominate the life sciences industry now compete by amassing huge property portfolios in the form of patented transgenic organisms, their constituent parts, and the technologies used to create them. Although the industry's use of blocking patents is not new, the application of this competitive strategy to life forms is. The cost of researching and developing new traits and organisms, soliciting patents and regulatory approval, paying licensing fees, and marketing new products has risen so high that few firms can afford to enter the industry.

The Waves of the Future

If the large-scale investments that the private and public sectors are making in these areas are any indication, genomics, bioinformatics, and proteomics are clearly the wave of the future. *Genomics* is the automated sequencing and analysis of genes; *bioinformatics* refers to the inference of genes' functions from information about known DNA sequences in other organisms; and *proteomics* refers to the science of protein functions and their relationship to genes. All three rely heavily on the new information technologies. The emergence of these new sciences, and their incorporation into plant and animal genetic-engineering research, has raised even higher the barriers to entry that characterize the agricultural biotechnology industry.

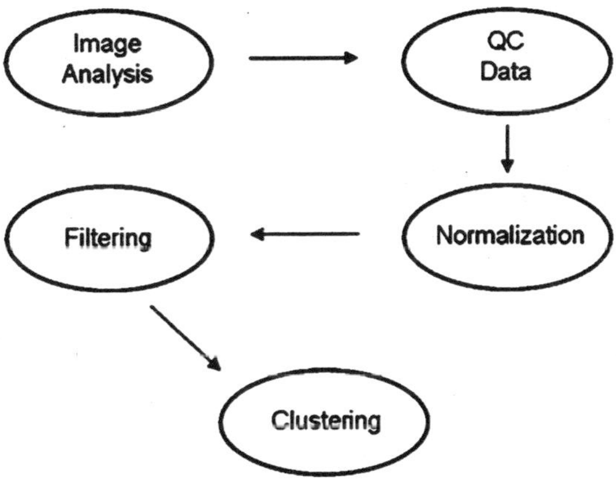

Fig. Comparative Genomics Pathway

William Boyd and Astrid Scholz (both in this volume) argue that genomics has become the driving force of innovation in the biotechnology industry, altering research agendas and augmenting competition over intellectual property rights. These new sciences promise tremendous acceleration of product research and development (R&D)—an all-important attribute in a field in which speed is the essence of competition. Indeed, the language of speed, racing, and getting there first are pervasive in corporate discussions and in descriptions of the industry. Genomics and bioinformatics have enabled firms to analyze many more organisms and traits at a much lower cost, greatly augmenting the number of patents for which firms can apply. The cost of "going genomic," however, is high, which is why most life sciences firms have entered into partnerships, joint ventures, and other strategic relationships with specialized genomics and bioinformatics firms.

The development of these new information sciences has had other consequences as well. The pharmaceutical industry's embrace of these new technologies in its drug discovery process over the last decade has had important implications for the industry's uses of genetic resources from the global South. As Astrid Scholz shows in this volume, the technological changes that have swept the industry have pushed the pharmaceutical sector toward combinatorial chemistry and "rational" drug design, reducing the economic significance of natural products.

As a result, the balance of power in genetic-resource negotiations has shifted even further toward pharmaceutical firms, which can now credibly claim that they are less in need of the South's genetic resources. In the future, this reduced dependence is likely to limit the economic benefits Southern countries can derive from their biological diversity.

THE NEW BIOTECHNOLOGY POLITICS

The spread of biotechnology; the consolidation of the biotechnology, seed, chemical, and pharmaceutical companies into a powerful life sciences industry; and the rush to patent genetic organisms and other constituent building blocks of life have all contributed to the recent politicization of the technology, an occurrence that none of the literature predicted. Indeed, the past five years have seen an immense escalation in political activity focused on the use of genetic engineering in agriculture. This activity is starting to shape the development of the technology and may affect it even more in the future. Any attempt to theorize the process of technological change must take this important influence into account.

An interesting question to ponder is why none of the scholars and analysts of biotechnology anticipated either the depth or power of the social response to the technology's development and deployment, or the political, economic, regulatory, and institutional crises this response would precipitate as the century came to a close. In similar fashion, the literature has also glossed over the real scientific challenges—what Andrew Pickering would call the "resistances of nature"—that have accompanied efforts to genetically engineer specific plants, animals, and marine organisms. Addressing the resistances of nature, although we do not do so here, is a critical next step for agro-food studies, as David Goodman persuasively argues in this volume.

Sources of Resistance

Although the most serious political challenges to the biotechnology industry have come in the past several years, social tensions about the new biotechnologies are neither new nor isolated from other social issues and conflicts. Ever since it became common knowledge that scientists *could* transfer genes across species in ways that were unlikely to occur in nature, individuals and groups have publicly questioned whether they (or we) *should* be doing so. Indeed, a close look at the North American antibiotechnology movement suggests that it has been around nearly as long as the technology.

Social resistance to the use of genetic engineering in agriculture has come from many quarters and is rooted in a broad array of moral, cultural, material, health, and environmental concerns. In Western Europe, where antibiotechnology activism has been most intense, sustained, and effective, activists' critiques have centered on the health and environmental implications of GEOs and the metabolic threats they represent to the living land and the living body. Derrick Purdue's analysis shows how European groups mounted a powerful challenge to biotechnology in the mid-

1990s, a challenge based on the mobilization of "counter expertise" that combined "a situated knowledge of seed science and law with impressive political skills".

Coterminous with other food and health scares such as mad cow disease and a growing anti-imperialist (and anti-American) sentiment, public opposition to biotechnology in many European Union (EU) countries has united consumers, farmers, environmentalists, and others in a fierce rejection of the technology and its purveyors. Social activists, particularly the environmental activists, have been instrumental in mobilizing this public opposition to biotechnology—most importantly, among consumers—by tapping into deeply held cultural and aesthetic values with respect to food and its production, and their connection to the preservation of (a simultaneously real and imagined) rural society.

Ethical concerns about whether human beings ought to be playing God like this, to paraphrase an oft-quoted comment of Prince Charles's, have also figured prominently in the rejection of the technology. In David Goodman's view, the activism against the new agricultural biotechnologies reflects a new kind of biopolitics that rejects the modern ontological division between nature and society and recognizes a "shared community" between human and nonhuman nature. Somewhat less sanguine about the anti-GE movement, Richard Lewontin sees it as embodying a powerful romanticism, in which activists' rejection of biotechnology is associated with a more general revulsion toward the exigencies of late capitalism. Although both views are partially right, they fail to reveal the full range and depth of sensibilities about genetic engineering.

Although no region has manifested a reaction quite as powerful or broad-based as that witnessed in Europe, the politics of agricultural biotechnology have been heating up in other parts of the world as well. In the United States and other advanced capitalist countries (e.g., Japan, Australia, and New Zealand), a broad set of activist groups and organizations has come together to protest agricultural biotechnology on many of the same grounds articulated in Europe. Compared with the uproar in Europe, however, social resistance to biotechnology in most other Northern countries is still significantly more activistbased than consumer-based, though the threat that the movement could have wider popular appeal presumably goes a long way toward explaining the industry's fierce resistance to GE food labeling.

In the global South, where political struggles over agricultural biotechnology have also grown in frequency and intensity, social resistance has revolved around a somewhat different set of issues. One is "biopiracy," or the appropriation of traditional knowledge and biological resources produced by generations of farmers, particularly indigenous communities. The notion that multinational corporations based in the North want to go to Southern countries to search for genetic material to take home, patent, and then resell to Southern farmers has reignited old resentments about North South economic inequalities and the history of external exploitation.

Concerns about biopiracy have been linked to a broader rejection of certain Northern countries' efforts to establish intellectual property rights (IPRs) over life forms, although the politics of IPRs are intrinsically complicated and reflect a wide diversity of perspectives on political-economic strategies for achieving community, local, and national economic and political empowerment. To many, the notion that life could be patented is absurd and accepting that notion represents a perilous path for indigenous communities and Southern countries to take, given existing world power inequalities. Others see the fight against the extension of IPRs as a losing battle. Because the advanced industrialized countries have patent rights to *their* agricultural commodities, they argue, Southern countries and communities should seek to establish such rights as well and should forge agreements that will allow them to maximize the benefits they receive from their genetic resources.

Two other issues—namely, threats to the environment and harmful effects on poor farmers—have galvanized social resistance to biotechnology in both North and South. Worries about the changes that GEOs could introduce into local ecosystems have figured prominently in national and international debates about the safety of GEOs, since many Southern countries are home to the wild relatives of GE species. Critics of the technology have drawn an analogy between the probable social and economic consequences of biotechnology and those of the Green Revolution, which the critics blame for increasing farmers' dependency on Northern corporations for inputs, for reducing the viability of poor farmers, and for harming indigenous communities.

However, the politics of biotechnology, the environment, and the fate of small farmers are far from clear or uncomplicated, since some Southern scientists, academics, and poor farmers see biotechnology in a positive light—as offering a way to lower the costs and minimize the environmental effects of a chemical-intensive agriculture while increasing the yields of local staple crops. Indeed, some Southern scientists have expressed considerable frustration at foreign and domestic nongovernmental organizations (NGOs) for hamstringing their efforts to develop agricultural biotechnologies that they believe could benefit their countries' poorest farmers.

Forms of Resistance and Arenas of Struggle

Political struggles over the new agricultural biotechnologies and their close cousins, genetic resources, are taking numerous forms and are being fought out on many levels and in multiple arenas. Although colourful protests and the destruction of GE test plots are the actions that have made the most headlines, contemporary biotechnology activism also encompasses lobbying efforts, litigation, regulatory struggles, public education and consciousness raising, and discursive struggles, as well as politically motivated consumer avoidance of GEOs, or what Melanie Dupuis describes as a "Not in My Body" reaction to genetically engineered foods.

In the advanced industrialized countries, some of the most serious political challenges have been mounted in the regulatory and policy spheres. Many Northern activist organizations have focused on food labeling, although as Julie Guthman shows, the politics of labeling are ambiguous, and labeling could easily lead to results that are both politically counterproductive and undesirable to many in the sustainable food movement. In the South, too, political struggles have involved legal and regulatory battles (in Brazil, India, and Chile, for example), though for Southern nations these have been rivaled in importance by activities in multilateral negotiating bodies and forums, such as the UN Convention on Biodiversity and the UN Cartagena Protocol on Biosafety.

Indeed, the various UN bodies and conventions have been critical arenas of struggle for Southern countries because they represent nodes of control for the transnational movement of germplasm and genetically engineered organisms. As a number of scholars have argued, the battles over biotechnology and biodiversity are as much about discourse— the power to name and define— as they are about material outcomes. Business firms, states, NGOs, and local communities are struggling not only over material access to and control over plant genetic resources but also over the meaning of terms such as *resources, nature,* and *the environment* more broadly and over whose value systems, worldviews, and knowledge will prevail when it comes to defining and valuing them.

Similarly, when activists and the biotechnology industry are at loggerheads over whether GE foods are safe for the human body and other biological systems—and over what kinds of scientific criteria should be relied upon to decide—these conflicts are simultaneously material and discursive. Indeed, in Sheldon Krimsky's view, discursive struggles are ultimately more important than the "surficial" struggles, for "those who eventually gain control over the bio-mythmaking will affect the pathways of innovation for future generations". Although control over mythmaking is probably best regarded as something that is dynamic and shifting rather than as something that can be conclusively gained or captured, Krimsky's more general point about the significance of discursive struggles is well taken.

Finally, although much of the social resistance to biotechnology is national, there is also an important transnational element to the movement. The nature and import of that transnationality needs to be more seriously investigated than it has been to date, however. (Indeed, most aspects of the biotechnology politics discussed here have not been adequately theorized.) Toward this goal, social organizing around biotechnology could be fruitfully compared with other transnational social movements, and their modes of organizing and their effectiveness could be analyzed.

Impacts of the New Biotechnology Politics

What effect has the explosion of GEO politics had on the biotechnology industry and on the development and deployment of the technology? Although the effects of biotechnology activism require much more analysis, the studies in this volume suggest that the groundswell of social resistance, both organized and unorganized, is profoundly affecting the industry and the trajectory of technological change. Social resistance has substantially reduced the market for GE products, particularly in Europe; has negatively altered the political climate surrounding the technology by raising concerns about its potential consequences; and has significantly dampened Wall Street's enthusiasm for biotechnology and slowed the industry's momentum. The rising social conflicts have also brought tremendous media and public attention to the issue. In the language of actor-network theory, these new political conflicts have opened up the formerly "black boxed" practices of conventional agriculture—which have now come to include modern methods of genetic engineering—and subjected them to intense scrutiny. In the process, private corporations' rights to introduce new technological innovations unimpeded have been challenged.

In our chapter on antibiotechnology activism, Munro and I argue that organized resistance to biotechnology has affected the fortunes of the biotechnology industry and the political environment surrounding the technology in three important ways: first, by pushing supportive states (including the United

States) to take more seriously the task of regulation; second, by helping to create new, supranational regulatory regimes for GEOs; and third, by imposing significant economic costs on the industry. In general, our work points to an emerging shift toward the re-regulation of the technology and industry, representing an important countertendency to the deregulatory trend of the past two decades.

Dennis Kelso's chapter on genetically engineered salmon demonstrates that the rising social resistance to biotechnology operates in more subtle ways as well. His empirical analysis shows how public perceptions of other GE foods are influencing salmon farmers' thoughts about whether they will adopt transgenic salmon grow-out stock, if and when that GE salmon is approved by government regulators. More generally, Kelso suggests that the commercial adoption of transgenic organisms is not preordained by the mere availability of this new technology and the competitive edge it may offer to some, but will instead depend upon social agency and struggle, including the raging discursive battles over the costs, risks, and benefits the technology promises to convey to different sectors of society and to the environment. Kelso's analysis also shows the real diversity of interests within the salmon sector, reminding us that no industry is totally unified in its interests or behaviour and that this diversity may strongly influence the process of technological change, as different groups decide whether or not to adopt the technology or sell products that are made with it.

Thus far, this discussion has focused on how social agency and activism are affecting the deployment of the new agricultural biotechnologies and the economic health of the agricultural biotechnology industry. But the new biotechnology politics will have effects far beyond these realms. As Frederick Buttel points out, the economic globalization regime is now rife with political tensions over GEOs. In his view, the politicization of agricultural biotechnology may turn out to be the Achilles' heel of the North's "globalization project, " that is, the organized attempt by a set of globally oriented state policymakers, financial and transnational corporate elites, and managers of global institutions (for example, the International Monetary Fund, World Bank, and World Trade Organization) to stabilize twenty-first-century capitalism. Another way in which biotechnology politics has been significant is through its impact on institutional change.

Institutional Restructuring

As a number of scholars of technology have noted, new technologies are not formed from whole cloth but rather emerge out of particular historical circumstances and existing institutional arrangements. *Institutional arrangements* refers to the institutions, organizations, international agreements, and bodies of national and international law and policy that structure the agro-food system, as well as to the informal norms, practices, customs, and power relations that pattern the everyday behaviour of relevant actors. Biotechnology's development has clearly been influenced by a long list of institutions, including prevailing property-rights regimes; the relationship between the public and private sectors (especially between universities and industry); scientific conventions, norms, and practices; institutionalized market relations between groups in the agro-food supply chain; the agencies and institutions that collectively create and constitute the regulatory environment; and trade relations and patterns. It is within this institutional context that biotechnology firms have devised and pursued their R&D strategies, that scientists have developed and tested new technologies and applications, that farmers have decided how and what to plant, and that consumers have made choices about what they are willing to eat.

Yet, the new biotechnologies not only have evolved out of a particular set of institutional arrangements but also act as springboards for new ones. Indeed, biotechnology— in the broadest sense of the word—serves as a catalyst for the formation of new institutional arrangements in society. Although its catalytic role cannot be seen in isolation from other political, economic, and ideological forces at work in late-twentieth-century capitalism—for example, the material and ideological changes associated with the ascendance of neoliberalism— biotechnology is playing a unique and provocative part in the process of institutional change.

The StarLink corn incident powerfully illustrates this observation. After it became clear that neither the relevant U. S. government agencies (the USD A, the EPA, and the Food and Drug Administration) nor the agro-food industry was capable of keeping the unapproved corn out of the human food supply, new norms for genetic testing, new protocols for growing and handling GE crops, new regulatory standards, new labeling requirements, and new systems of distribution were developed. In general, the stresses generated by biotechnology created demands for the institutional reforms that are now being pursued.

New Governance Structures

Examples of the institutional changes provoked by the scientific, social, economic, and environmental aspects of the biotechnology revolution are many and range from relatively insignificant changes to profound changes in the social norms, relations, and governance structures characterizing the agro-food system. Perhaps the most obvious arena in which biotechnology has generated a demand for institutional change is that of intellectual property rights. As the life sciences firms have sought to capitalize on their enormous investments in genetic engineering, genomics, and drug discovery and development, they have demanded greater protection of private property rights. The need to establish proprietary rights over living organisms is seen as paramount, since such rights remain the primary means by which firms can profit from their investments in research, development, and acquisition.

This push for institutional change in the intellectual-property sphere is not a new phenomenon; nor is it associated exclusively with modern biotechnology or with the biotechnology industry. As Jack Kloppenburg showed in *First the Seed,* private plant-breeders in the United States and Europe have long sought to establish proprietary rights to plant genetic resources and have been quite successful. Yet some features of the present period—and of the biotechnology industry's involvement with this effort are particularly notable.

For example, efforts to extend IPRs to life forms are now being carried out at the *international* level as well as the national level. In response to the concerns of the biotechnology industry (as well as other key industries), the United States, supported by other advanced industrialized countries, is pressuring other countries to adopt systems of private property rights that are commensurate with its own. These systems will extend protection for the new seed, plant, animal, and process technologies into developing countries, most of which have patent, copyright, trademark, and trade-secret protections that are far less comprehensive than those in the United States. The primary (but by no means the only) vehicle for this effort is the World Trade Organization (WTO). During the Uruguay Round of the GATT negotiations (the precursor to the WTO), the advanced industrialized countries pushed for the Agreement on Trade Related Aspects of Intellectual Property Rights (TRIPS), which will impose minimum IPR standards on all countries by 2006. To remain in compliance with the WTO, countries will have to change their national laws to meet these standards.

A second noteworthy change is that whereas in the past it was primarily independent seed companies that were pushing for proprietary rights in germplasm, it is now a vertically integrated and economically powerful life sciences industry that is pushing for proprietary rights with regard to a much broader array of life forms. The marriage of the agricultural biotechnology, chemical, pharmaceutical, and seed industries has created a coalition whose interests in IPRs are similar and who collectively wield tremendous political power. The progress made in recent decades in extending private property rights to life forms testifies to the power of the life sciences sector to shape the institutional environment in its favour.

The public's reaction to biotechnology is significantly changing two other institutions: the food-supply and food-distribution systems. The rejection of GEOs by European consumers, along with the fallout from the StarLink debacle, is stimulating the development of a new "identity preservation" system that will eventually make it possible, at least in theory, to segregate genetically engineered crops from non-engineered crops. Although it is not clear that such segregation is feasible in practice—particularly in open-pollinated crops, such as corn and canola—new storage facilities, farm management practices, transport and processing systems, and even easier-to-clean machinery are being developed. These changes will extend and complicate the commodity supply chain as existing industry actors alter their crop-growing and crop-handling methods, as new businesses offer new services to grain growers and suppliers (e.g., genetic testing, certification, and separate storage systems and facilities), and as specialized markets are established.

Another sort of institutional change is reflected in the plethora of initiatives taken at the local, regional, national, and international levels to deal with the economic, environmental, and human-health concerns raised by this new technology. Interestingly, much of this change is about (re)regulation and undoubtedly reflects as much a reaction to two decades' worth of neoliberal policy reforms and national de-regulation as it does the challenges posed by the new biotechnologies. In the United States, some localities have passed ordinances that ban or require labeling for GE food distributed in public programs, and states have considered (and in some cases passed) bills that would place temporary moratoriums on the planting of GE crops and make it illegal to raise GE fish in lakes and ponds connected to other waterways.

Although most of these efforts represent small and highly circumscribed challenges to the biotechnology industry, they do represent institutional change. Much more sweeping, however, are the changes that have taken place in national and transnational regulation of GEOs. For example, the growing demand for GEO-free products in many countries has created significant pressure for national labeling systems that institutionalize this aspect of postproduction "choice". At the international level, the growth of trade in GEOs has led to the Cartagena Protocol on Biosafety, an international agreement to regulate the international movement of GEOs, and the establishment of an intergovernmental task force on genetically engineered foods within the Codex Alimentarius Commission, the UN organization that sets international food and safety standards. Within the Codex there is considerable pressure (particularly from Europe and Japan) to establish a system that will allow GE products to be traced from field to market. The establishment of such a system would represent a major shift in the way foods are identified, labeled, and internationally traded.

The Process of Institutional Change

Although little theorizing has been done about these processes of institutional change, many of the reforms spurred by biotechnology have clearly evolved out of existing institutions that have been modified to fit new circumstances or extended into new spheres, or both. For example, as Scott Prudham shows in his chapter in this volume, the genetic-engineering research now being done on trees in the U. S. Pacific Northwest reflects a long history of public-sector involvement in and support for research aimed at increasing commodity production. Yet the institution of the silviculture research cooperative itself is changing, both in response to IPR developments and to changes in the broader political economy of public science, which have made public funds more difficult to come by while offering new opportunities for the commercialization of publicly supported research through the Bayh-Dole Act of 1980.

Although the future of such cooperatives is difficult to discern, the old model of cooperation in tree improvement and the free sharing of plant genetic material among co-op members is likely to be supplanted by new models of academic-industrial relations that "emphasize more-formal management of research as intellectual property ... [and] more-exclusive relationships, public and private".

Historical continuities are also evident in the area of intellectual property rights, which, as already noted, have been a long-term concern of private seed companies in the United States and Europe. In this case, the efforts of the biotechnology industry and the governments of advanced industrialized countries to create more comprehensive—and increasingly global—IPR systems can be seen as an attempt to extend the authority of an institution into new geographical and biological spheres. In the process, however, such institutions tend to change significantly as they are adapted to new circumstances and are reshaped by local social and environmental forces. We need to study both aspects—efforts to extend particular institutions, and the process of transformation by local actors—if we want to understand processes of institutional change.

Although biotechnology's coming of age has stimulated the transformation of existing institutions, wholly new institutions, patterns, and relations are also being established. As Scholz describes in her chapter, the demand for genetic resources by the large pharmaceutical companies has generated new institutional arrangements in the form of bioprospecting fees paid to various state and nongovernmental institutions. These fees follow certain norms established by pioneering agreements such as that between Merck and Company and the Costa Rican biodiversity institute named INBio. The Cartagena Protocol discussed in several chapters of this book also had no historical precursor and was a direct outcome of the new environmental challenges raised by biotechnology and concerns about its deployment on a world scale.

Tremendous political struggle characterizes the process of institutional change because particular groups seek to promote, protect, and institutionalize their interests. These political struggles are being fought in a variety of settings, from the informal, grassroots, and local, to the formal, national, and international. But it is not only their ubiquity and diversity that should command our attention; so too should the fact that the process of struggle itself can lead to new and unanticipated agreements and alliances between stakeholders and, in some cases, to unexpected institutional outcomes.

Yet, while struggles over institutional reform invariably reflect existing power dynamics and relations, they do not necessarily (or perfectly) reproduce them, because different groups may alter or crystallize their definitions of self-interest, establish new alliances, and enter into new political arrangements and compromises. Hence, the very process of political struggle can create opportunities for altering power dynamics and relations in the future through the re-evaluation of existing patterns and the establishment of new norms, regulatory frameworks, and institutional relationships.

THE BIOTECHNOLOGY MILLENNIUM

I close with three final observations about biotechnology, its meaning, and its future. First, many of the technological, social, and institutional developments that are occurring around biotechnology today are fundamentally ambiguous. As the most astute students of technology have argued, technologies embody emancipatory as well as oppressive potential; indeed, the meaning, consequences, and transformative potential of any particular technology depend upon how that technology is deployed, by whom and for what purposes and upon the meanings it is given by those who use it. In the context of the technologies that concerns us here—those involving the genetic engineering of plants, animals, and other living organisms—it is not hard to imagine liberatory and positive possibilities, as well as the more negative scenarios painted by biotechnology's critics.

A second and closely related observation is that social, historical, and environmental context matters. Who makes decisions about the development and deployment of a particular technology, where and how it is deployed, how much control various social groups have over these decisions, and how the nonhuman world acts and reacts—all these contextual factors matter enormously in determining the myriad consequences of technological change. This is the reason that it is so important to investigate and analyze the patterns of control and decision making in the case of contemporary agricultural and other related biotechnologies.

The final observation has to do with the broader political-economic developments that are taking place in the biotechnology field and that are poised to influence profoundly the contexts just described. On the one hand, we are clearly in the midst of a major restructuring of the agro-food system, a process to which biotechnology is an important contributor. Large corporations seem to be gaining greater control over many factors: the technologies used to produce food, fibers, and raw materials; the genetic material on which they are based; the labour, land, and production practices of farmers; and the food-distribution and food-marketing systems. On the other hand, and before we conclude that hegemonic corporate control over the agro-food system is the only possible future, we must recognize that no single set of actors—be they life sciences corporations, agro-industry, transnational elites, or powerful states—fully controls the agro-food system.

Even though the multinational life sciences corporations appear to have amassed unprecedented power during the last decade and half, that power is contested by a wide range of groups and actor networks. Indeed, the very process of moving specific biotechnologies from inception to commercialization has brought them into the public eye and given civil society access to them, in test plots and production practices as well as in the marketplace. These technologies' very presence in such public places makes them vulnerable to challenge and to alternative uses and interpretations. The politicization of biotechnology could have a highly democratizing effect on the agro-food system. Or at least these struggles will help to form the new political terrains of the biotechnology millennium.

Chapter 11

Assessing the Consequences of Genetic Engineering

Assuming that there are no intrinsic ethical grounds for condemning biotechnology and genetic engineering, the moral status of these practices can only be determined by an evaluation and assessment of their expected consequences. In theory, such an assessment should not be too difficult to carry out, since the advantages and disadvantages of gene-splicing techniques are fairly well publicized. in practice, however, the situation is different. The actual consequences of applying advanced molecular biology depend upon the social and political setting in which the application takes place. As different groups of people hold different views concerning the structure and dynamics of social and political life, these groups inevitably also disagree upon the consequences of the development of recombinant DNA techniques.

My aim in this paper is to soft out and analyse, first, the major consequences of genetic engineering, and second, the most important ideological factors which cause dispute concerning their weight. I shall begin by reviewing the advantages and disadvantages of various forms of biotechnology as they have been presented by the proponents and opponents of these practices. I shall then go on to consider the different ways in which the effects of genetic engineering can be evaluated, depending on the degree of optimism or pessimism inherent in the referee's judgment.

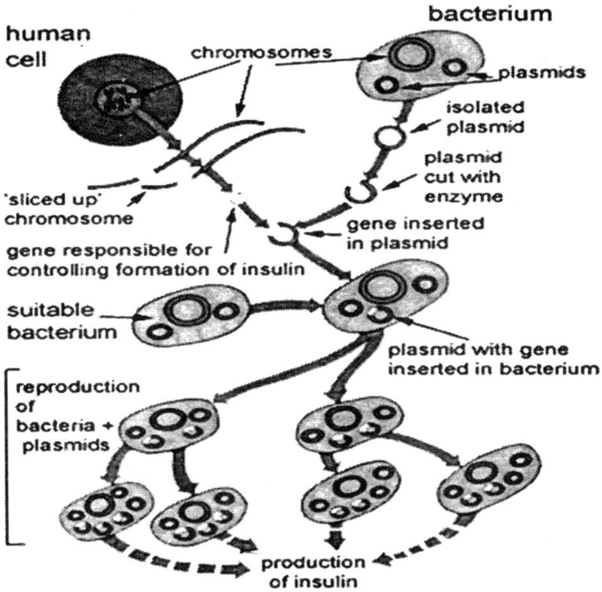

Fig. Genetic Engineering

THE ADVANTAGES OF GENETIC ENGINEERING

The advantages of biotechnology, as seen by its proponents, include many actual and potential contributions to medicine, pharmacy, agriculture, the food industry, and the preservation of our natural environment.

Within medicine and health care, the most far-reaching consequences are supposed to follow from the mapping of the human genome, an enterprise which is still in its infancy but which holds the key to many further developments. Accurate genetic knowledge is a precondition for many foreseeable improvements in diagnoses and therapies as well as an important factor in the prevention of hereditary diseases, and possibly in the general genetic improvement of humankind. Provided that such knowledge becomes available, potential parents can in the future be thoroughly screened for defective genes and, depending on the results, they can be advised against having their own genetic offspring, or they can be informed about the benefits of pre-natal diagnosis.

One of these benefits is that an adequate diagnosis may indicate a simple monogenic disease in the foetus which can be cured by somatic cell therapy at any time during the individual's life. Another possibility is that an early diagnosis reveals a more complex disorder which can be cured by subjecting the embryo to germ-line gene therapy-this form of prenatal treatment also means that the disorder will not be passed down to the offspring of the individual. Even in the case of incorrigible genetic defects the knowledge benefits the potential parents in that they can form an informed choice between selective abortion and deliberately bringing the defective child into existence.

Pre-natal check-ups and therapies do not by any means exhaust the medical applications of future genetic engineering. Somatic cell therapies are expected to help adult patients who suffer from monogenic hereditary diseases, and the increased risk of certain poly genie diseases can be counteracted by providing health education to those individuals who are in a high-risk t racket. In addition, the purely medical benefits of genetic engineering are extended by the fact that gene mapping will probably prove to be useful to employers and insurance companies as well as to individual citizens. Costly mistakes in employment and insurance policies could be avoided by carefully examining applicants' tendencies towards illness and premature death prior to making the final decisions.

Gene therapies and genetic counselling are practices which require advanced knowledge concerning the human gene structure. But the genetic manipulation of non-human organisms can also be employed to benefit humankind. The applications of gene splicing to pharmacy, for instance, may in the future produce new diagnosing methods, vaccines and drugs for diseases which have to date been incurable, such as cancer and AIDS. The agricultural uses of biotechnology include the development of plants which contain their own pesticides. In dairy farming, genetically engineered cows give more milk than ordinary ones, and with the right kind of manipulation the proteins of the milk can be made to agree with the digestive system of those suffering from lactose intolerance. As for other food products, biotechnology can be applied to manufacture substances like vanilla, cocoa, coconut oil, palm oil and sugar substitutes. And biotechnology can even provide an answer to the growing environmental problems: genetically engineered bacteria can be used to neutralize toxic chemicals and other kinds of industrial and urban waste.

THE DISADVANTAGES OF GENETIC ENGINEERING

The disadvantages of biotechnology, as seen by its opponents, are in many cases closely connected with the alleged benefits. An efficient strategy in opposing genetic engineering is to draw attention to the cost and risk factors which are attached to almost all inventions and developments in the field.

One general critique can be launched by noting that the applications of recombinant DNA techniques are enormously expensive. Millions of pounds are spent every year by governments and multinational corporations in biotechnological research and development. These resources, opponents of the techniques argue, would do more good to humankind if they were allocated, for instance, to international aid to the Third World.

Another problem is that, despite the undoubtedly good intentions of the scientists, the actual applications of genetic engineering are often positively dangerous. Consider the case of plants which are inherently resistant to diseases, or which contain their own pesticides. Although there are no theoretical obstacles to the production of such highly desirable entities, corporations-who also sell chemical pesticides-might prefer to market another type of genetically manipulated plant, which is unprotected against pests but highly tolerant to toxic chemicals. The result of this policy would be an increase in the use of dangerous chemicals in agriculture, particularly in the Third World-which is to say that the outcome is exactly opposite to the one predicted by the proponents of biotechnology. Besides, it is quite possible that genetically engineered grains are less nourishing than the grains which are presently grown. If this turns out to be the case, then the employment of biotechnology will intensify instead of alleviating famine in the Third World. And to top it all, genetically manipulated plants may contain carcinogenic agents, and thus contribute to the cancer rates of the developing countries.

An oft-used criticism against agricultural biotechnology is that the introduction of altered organisms into the natural environment can lead to ecological catastrophes. Scientists working in the field of applied biology have themselves noticed this danger, and set for themselves ethical guidelines which are designed, among other things, to minimize this particular risk. But as the opponents of genetic engineering have repeatedly pointed out, not all research teams follow ethical guidelines if the alternative is considerable financial profit.

Apart from the excessive expense and the increased risk of physical danger caused by biotechnology, the opponents of gene splicing can appeal to yet another disadvantage, which is related to widely shared moral ideals rather than to straightforward estimates concerning efficiency. Genetic engineering, it can be argued, is conducive to economic inequity and social injustice, both nationally and globally. Even in the most affluent western societies, gene therapies are too expensive to be extended to members of all classes and age groups. Subsequently, these therapies are likely to become the privilege of an elite, and they will drain scarce resources from the more basic areas of health-care provision. In the developing countries, the situation is even more absurd. Medical problems which originally stem from lack of democracy, lack of education, shortages of fresh water, population explosion, archaic landownership arrangements and the like cannot possibly be solved by high-tech western innovations which can barely be made to work in the most affluent and democratic of countries.

Another type of injustice emerges from the fact that the natural national products of many developing countries are superseded in the market by the biotechnological products of multinational corporations. Genetically engineered substitutes for sugar, for instance, could adversely affect the lives of nearly fifty million sugar workers in the Third World. Biotechnological vanilla could increase the unemployment figures by thousands in Madagascar, Reunion, the Comoro Islands and Indonesia. And plans to produce cocoa by genetically manipulating palm oil threaten the current export market of three poverty-stricken African countries, namely Ghana, Cameroon and the Ivory Coast. In all these cases, the profits of multinational Western corporations are clearly and directly drawn from the national income of the developing countries.

Finally, as regards medical biotechnology, there are those who believe that advanced knowledge concerning the human genome will inevitably become an instrument of genetic programming, which in its turn leads to subtle forms of genocide and general injustice. The opponents of gene splicing argue that the development begins inconspicuously with attempts to eliminate hereditary diseases. This practice of what is called 'negative' eugenics will, however, soon be followed by more 'positive' efforts towards altering the human genome: the inborn qualities of future individuals will first be improved in their own alleged interest, and then, later on, in the interest of society at large. When this development has gone far enough, scientists will also be asked to design special classes of sub-human beings who can do all those occupations which are too dangerous or too tedious for ordinary people. The outcome, according to the opponents of biotechnology, will be something like Aldous Huxley's Brave New World.

Technological Determinism

When one compares the views for and against biotechnology, one can easily see that the prevailing differences of opinion are not purely factual in nature. It is an undeniable fact that genetic engineering can be employed to eliminate diseases, but it is also an undeniable fact that genetic engineering can create bizarre and dangerous life forms. Likewise, it is possible that agricultural biotechnology will relieve famine in the Third World, but it is also possible that it will intensify it.

One ideological difference between those who defend recombinant DNA techniques and those who oppose them is based on their attitudes towards what have been called in the literature 'technological imperatives'. The proponents of genetic engineering tacitly assume that the development and implementation of new techniques can always in the end be controlled and steered by human decision-making. If this were the case, then there would be no binding imperatives dictated to us by technological processes. This view has been labelled in the literature 'technological voluntarism'. On the other hand, however, the opponents of genetic engineering have argued that technological development in fact has its own internal laws and logic, which cannot be altered or checked by human choices or human action. This view, which can be called 'technological determinism', implies, among other things, that all new techniques, however dangerous or morally offensive they may be, will ultimately be implemented.

Those who believe in technological determinism can support their views by referring to real-life examples drawn from the history of science and technology. The splitting of the atom, for instance, made it possible to devise nuclear explosives, and despite the expected evil consequences, the atom bomb was constructed and used as soon as the technical difficulties had been solved. In medicine, the drastic improvement of life-sustaining therapies since the 1950s has led to a situation where irreversibly unconscious and intolerably painful human lives are prolonged beyond all reasonable limits, the only boundaries being set by medical technology. These and many other examples seem to show that every technological possibility will ultimately be exploited whatever the effects on human well-being.

Those who place their trust in technological voluntarism like to point out, however, that the 'imperatives' created by scientific know-how can always be rejected. Atom bombs were not launched by abstract technological systems, but by political decision-makers. Hospital patients are not kept alive against their own wishes and beyond any hope of recovery by medical technologies as such, but by doctors and medical teams whose actions are sanctioned by democratically chosen legislators. Even if it is true that technical solutions often 'offer themselves' to those making important decisions, these 'offers' can be ignored or rejected if the consequences of accepting them seem to be harmful or otherwise undesirable. The fact that technological imperatives can be disobeyed makes it possible, according to the voluntarist, for us to reject those forms of technology which are harmful while preserving those forms which are useful to humankind.

Both technological determinism and technological voluntarism appear to have some truth in them. There are cases in which humankind seems to be guided, to a certain extent at least, by scientific and technical inventions. But there are also cases in which scientific knowledge and its applications are more or less firmly in our control. To name only one example of each class, the atom bomb was built and used in the Second World War, whereas the development and use of poison gases was foregone by the political and military decision-makers of the period. These and many other real-life examples indicate that the dispute between determinism and voluntarism cannot be solved at a general level. But they also indicate that technologies can be divided into those which support the determinist interpretation and those which uphold the voluntarist view.

Given that technologies differ from each other in this respect, the important question in our present context concerns the category into which genetic engineering can be placed. Is genetic engineering one of those technological advances which control our actions, or does it belongs to the class of techniques that we can control for our own benefit?

The empirical evidence which is presently available is mostly reassuring, and allows for moderate optimism concerning the controllability of biotechnology. Although there have been a number of unauthorized and potentially dangerous experiments with recombinant DNA, the guidelines set first by the scientists themselves and later on by governmental bodies have been rather leniently followed by the majority of research teams. During the last two decades, gene splicing has been transformed from invention to industry, but up to this time human decisions have definitely played an important role in the development and application of the technique.

The Three Stages of Technological Development

Granted that biotechnology has not yet taken total control over human affairs, the pessimist can nevertheless argue that the situation will in the near future be drastically different. The pessimist's view can be based on the idea that new techniques develop gradually, each stage provoking its own peculiar responses in the business world and among the general public.

The first stage of technological development is the theoretical invention of a new technical tool or process. Depending on the nature of the invention, this stage can either go unobserved by all but a few experts in the field, or it can become widely publicized and commented upon. The latter is often the case when the new technique in question is connected, however remotely, to human sexuality and reproduction. At the second stage of technological development the newly invented tools and processes are transformed by practical innovations into industrial products and means of production. Since this phase usually extends over a relatively long period of time, public attention tends to fade away, along with the fears which may initially have been caused by the novel device. At the third and final stage the completed products are marketed and distributed among the consumers. Assuming that the preceding innovation phase has been long enough, the products often enter the market without protestation, regardless of the reactions which may have accompanied the introduction of the new technique.

The three-stage model of technological development can be employed by the opponents of genetic engineering to construct an argument against scientific and industrial gene splicing. When recombinant DNA techniques were first invented in the 1970s, their potential applications were publicly discussed and widely condemned. The reaction was for the most part based on religious taboos and on horror images created by twentieth-century science fiction, not on any real-life facts. The practical outcome of the protest was, however, that scientists between themselves decided to proceed cautiously, and to avoid arousing further popular outrage. Under the cover of professional ethical codes, biotechnology was then gradually advanced to its present innovational stage, at which the actual and potential industrial applications have become numerous and diverse. Attitudes towards gene splicing have, in the course of this development, inconspicuously become more permissive, largely because of the fact that there have been no major catastrophes.

According to technological pessimists, the relaxation of popular attitudes marks the beginning of a new and increasingly dangerous era in genetic engineering. When scientists are no longer censoriously controlled by the general public, their readiness to follow voluntarily assumed guidelines will decrease, and their codes will be gradually slackened. At first the reasserted freedom of the scientists will be qualified by assessments of the expected harms and benefits. Although the majority of gene-splicing activities will be sanctioned, those forms of genetic engineering which are considered to be particularly hazardous will be kept under the closest surveillance. After a while, however, the qualifications will be removed. By the time biotechnological development fully reaches its third stage-the stage at which completed products are marketed and distributed- consumers will have become accustomed to whatever forms of genetic engineering scientists care to pursue. And at that point, so the opponents of biotechnology claim, humankind will only be one step away from the Brave New World.

There are two reasons to doubt the validity of the pessimistic view. First, as regards empirical facts, most western countries will during the 1990s enter a phase where genetic engineering becomes regulated by law. The emergence of legal regulations undermines the immediate importance of public attitudes and voluntary guidelines, and the progress from qualified to unqualified licence predicted by the pessimist cannot take place. Second, even if the pessimist's prediction turned out to be correct, it remains unclear whether the unrestricted practice of gene splicing would in fact lead to the kind of genetic totalitarianism that Huxley described. One cannot reasonably assume that biotechnology could by itself destroy the prevailing democratic institutions and replace them by an authoritarian system. It is true that decision-making concerning the development of genetic engineering is at present controlled by multinational corporations rather than democratically elected governments. But it does not follow from this fact that humankind would be on the path to genocide and totalitarianism.

It seems to me that there are two types of pessimism involved in the issue, one of which is more readily justifiable than the other. The first, legitimate type of pessimism concerns the power of biotechnology to solve global problems like famine and pollution. Since it is not in the interest of multinational corporations to organize international aid to Third World countries, it is indeed unlikely that suffering would in the near future be effectively alleviated by implementing gene-splicing techniques. But this is not to say that genetic engineering ought to be banned, or that the situation could not be different sometime in the future. The second, illegitimate type of pessimism cynically assumes that nothing can be done to change the situation, and that the total prohibition of gene-splicing activities is the only way to save humankind from the slippery slope to which mad scientists and big corporations are leading us. This attitude is groundless and, moreover, positively dangerous in that the prophesies of the pessimist of this second type can easily become self-fulfilling. If ordinary citizens do not even attempt to control the corporation giants to make the benefits of biotechnology available to everybody, it is obvious that they will not attain such control. Pessimism in this sense is a form of ideological defeatism rather than considered intellectual scepticism.

The Consequences

To return to the consequences of genetic engineering and their assessment, it is now easy to see why competing views on biotechnology are so often so uncompromisingly antagonistic to each other. Factual disputes, which could be reconciled by empirical observation, play a relatively minor part in the controversy, and the real disagreement can be traced to commitments to optimism or pessimism, voluntarism or determinism. It seems, moreover, that these commitments are unavoidable, i.e. that it is impossible to produce a genuinely impartial assessment of the consequences of genetic engineering. It is therefore necessary to assess the validity of the underlying views before an evaluation of biotechnology can be completed.

As regards technological voluntarism and technological determinism, the empirical evidence which is available at present seems to indicate that genetic engineering is, by and large, controllable by human decision-making. The temptations created by technological possibilities should not be ignored or mitigated, but they should not be exaggerated either. Like all advanced technological innovations, gene splicing offers many possibilities, some of which will appeal to certain unscrupulous research groups despite the inherent hazards and despite the voluntary codes agreed upon by more responsible scientists. But the legal regulation of biotechnology, which in most western countries will take effect during the 1990s, provides an adequate device against such unethical conduct.

The issue of optimism and pessimism is slightly more complex, since different variations of both views can be either acceptable or unacceptable in the context of biotechnology. To begin with the gloomier views, epistemic pessimism is justifiable when it prevents people from believing uncritically what genetic engineers tell them about the advantages of the new techniques. Graver types of ideological pessimism cannot, however, be condoned, since they are prone to be self-fulfilling: widespread resignation and defeatism is likely to produce consequences which would never have occurred if people had been more hopeful. Similar observations can be made about the variants of optimism. It would be rather naive to believe that the multinational corporations which presently control biotechnological development would voluntarily undertake to further general welfare and global justice. Epistemic optimism in this sense is therefore unwarranted. But it is not necessarily naive to believe that changes are possible in economic and political life, and that genetic engineering can in the future serve wider interests than it does at the present time. Ideological optimism is in fact supported by the feature which can be used to disqualify its pessimistic counterpart: if the belief in a better world is in any way self fulfilling, it is obviously better to hope and prosper than to despair and perish.

Assuming that democratic decision-making can be gradually extended to biotechnology, and assuming that genetic engineers continue to be controlled by voluntary agreements or legal regulations, the consequences of gene splicing can be expected to be prima facie good and desirable. Democracy both within and outside the world of biotechnology guarantees that no Brave New Worlds will be created as a result of misguided social experiments or malevolent dictatorial schemes. Surveillance and regulation, in their turn, provide a safeguard against reckless and inhumane research on areas such as biological warfare, the manipulation of our natural environment, and designs for new breeds or hybrids of animals. Granted that these possibilities are excluded, the remaining forms of genetic engineering are without doubt prima facie acceptable.

There are, however, certain embarrassing points which ought to be accounted for before recombinant DNA techniques can be wholeheartedly condoned. The moral problem at the core of the issue is that whatever advantages may ensue from the employment of genetic engineering, these advantages will not be allocated to those who need them most. The medical advances will almost exclusively benefit the wealthy and the privileged of the affluent western societies, and the industrial applications will mainly profit the big corporations, sometimes at the expense of destitute Third World countries. The prima facie desirable consequences of biotechnology are not genuinely worth pursuing, one might argue, unless they can be equitably distributed among nations and among individuals.

But although it is easy to point out instances of injustice in the application of gene splicing, none of these inequities is primarily caused by genetic engineering. The allocation of scarce resources is a growing problem in medicine and health care, but it has not been either originated or, to any significant degree, aggravated by the development of biotechnology. Gene therapies are not the only form of treatment which has been confined to the select few in our present-day societies. And although it is true that certain Third World countries will be even worse off than they are at present after multinational corporations have taken over their export markets, the roots of their suffering surely lie deeper. The invention of recombinant DNA techniques is, after all, a relatively minor evil compared to the centuries of western colonialism and imperialism which have left the national economies of these countries exceedingly vulnerable to any sudden changes in the world market.

What these remarks amount to is that genetic engineering should not be blamed for the injustice related to its implementation. If humankind can find a way to solve the problems of just distribution at a general level, then the benefits of biotechnology will also be distributed justly among individuals and nations. According to the moderately optimistic view that I have put forward and defended here, our best chance is to believe that this is possible, and then proceed to make it possible by our own actions. If this can be done, and if the requirements of democracy and adequate control are fulfilled, then the conclusion can be drawn that, in the last analysis, the expected consequences of genetic engineering favour its acceptance.

Chapter 12

Genetic Engineering in Theology

THEOLOGICAL ETHICS

My purpose in this essay is to give some account of the ways in which Christian theology and Christian ethics try to contribute to the debate about genetic engineering. There is in fact a large corpus of material on this topic, but some of it is repetitive. For the most part I shall analyse and quote from two authors and from a small number of 'church reports' which effectively point both to divisions of opinion and to consensus about resources and methodology. I am well aware that more, or different, reports could have been called in evidence. But, working here on a small scale, I think is sensible to confine myself in the way in which I have done. Moreover, there are a limited number of theological-ethical arguments and a limited number of methodological options, even if they come in different guises.

PAUL RAMSEY

Paul Ramsey was the doyen of American Protestant medical ethics over many years. His entry into this sphere of medical-ethical debate came as early as 1956, with an article called 'Freedom and responsibility in medical and sex ethics: a Protestant view', which was a critique of Joseph Fletcher's *Morals and Medicine* of 1954. For present purposes I quote from an article published by Ramsey in 1972. In this article two themes recur. These are the absolute ethical impropriety of embryo experimentation and the accusation that in the new bioethics, the experts are 'playing God'. In the article Ramsey is mainly dealing with IVF.

But the discussion is immediately relevant on both themes to the genetic realm. Ramsey had already developed his thinking on the latter theme in his *Fabricated Man: the Ethics of Genetic Control*. In the present context I confine myself to the charge of bioethics 'playing God'. Thus Ramsey refers to the IVF scientists who must 'mimic nature perfectly' and to 'this artificial mimicry of nature'. Again, 'we can clearly see the extent to which human procreation has already been replaced by the idea of "manufacturing" our progeny' and 'instead, the child as a product of technology is to be brought forth'. I treat Ramsey as standing at one end of a range extending from 'unpermissive' to 'permissive'. Joseph Fletcher represents the latter. Now I explore four reports in the middle part of the range.

Ethical Issues in the Biological Manipulation of Life

The above phrase is taken from the Report of the World Council of Churches 1979 Conference on 'Faith, Science and the Future' held at MIT, Cambridge, Massachusetts.

MIT followed the activist 1968 Uppsala WCC Assembly, which in turn followed the explosive 1966 Geneva Conference, which introduced revolutionary politics and the critique of technology on to the main WCC agenda-hence the keywords in the title of the Report are 'faith', 'science' and 'unjust'. So a strongly political perspective serves as a presupposition for the ethics of genetic engineering.

In considering certain biological, genetic and medical manipulations of human beings...we must keep sharply in mind the tragic consequences suffered by many people as a result of past eugenic theories and practices. These have all too often been used by groups in power against exploited or disenfranchised minorities ...[Hence,] the closest scrutiny of the social and economic conditions of...[human genetic analysis] application will be continuously required to ensure that such technologies contribute to a just and participatory society.

MIT wants at least to keep in balance both faith and praxis. Thus, theologically, 'as Christians we *believe* that we are both creatures of God and co-creators with him in fulfilling the image He has given us. This *belief forms* our high estimate of human worth.' That *belief* is then immediately linked with *action*. 'We are also guided by the words of Jesus: "Inasmuch as you *do* it unto one of the least of these you *do* it unto me."' *MIT is* the most politically and socially selfconscious of the texts which I examine.

MIT observes a sharp ethical distinction between somatic and germ-line genetic engineering. 'An individual may give informed consent to a change in his own body as in body cell alteration... and that change is not transmitted to future generations. *The individual cannot give informed consent to the genetic alteration of subsequent generations.*' Such decisions should be taken in a much broader social context. They cannot be just the products of contracts between scientist and subject.

So, among 'issues of grave concern' is gene replacement. Some features of this can only be ascertained by experimentation, 'first on animals over several generations but eventually on human subjects'. Echoing Ramsey, *MIT* concludes that 'someone has to be the first human experimental subject'.

Critical remarks about theology are not confined to *MIT*. But *MIT* forcefully claims that 'our understanding of the relationship of God to human life and non-human life requires much further development if we are to have a substantial ethical and theological base for evaluating genetic engineering in humans and other vertebrates. In our judgement, *theology does not yet speak in a meaninful way to the problems raised by contemporary biology*'. The meaning of this passage is by no means clear. Partly, it may mean that theology knows how to talk about human beings, but not about non-human, impersonal natural processes.

Experiments with Man

Experiments with Man: Report of an Ecumenical Consultation (EWM), edited by Hans-Ruedi Weber, was published in 1969 by the World Council of Churches and Friendship Press, New York. It is divided into three parts-'A record of convictions and questions', 'Working papers', and 'Meditations on man and his vocation'.

'One aspect of human existence needs close scrutiny now, namely man's power to discover, control, consciously influence and change his environment'. This raises profound questions. Can humans, in fact, influence and change nature and history? What is the purpose of such human dominion over nature and history? What do humans do with their dangerous knowledge? Are there any limits to human power? *EWM* recognizes that there are searching questions to be *put prior* to addressing biotechnological issues.

There are four types of experiment:

1. Routine treatment which, with each patient, is something of an experimental venture.
2. In the course of treatment, data are collected by which existing hypotheses, methods and medicaments are checked and improved and new ones developed.
3. Fundamental clinical research not primarily directed to any immediate therapeutic benefit.
4. There is biomedical research relating to the manipulation of genetic material. Here *EWM*, from 1969, peers forward and remarks that the ethical considerations attending research of this sort would differ markedly from those proper to the other three types. *Here, the investigator and society mast somehow act as guarantors of the integrity of future generations by defining the limits permissible in the manipulation of the basic material of the human species.*

'Defining the limits' becomes a recurrent theme in these texts.

EWM is strong on method. It has a useful note on natural law to which I shall refer later. It also goes further than *MIT* in making clear its expectations from theology:

Today the old theological questions [set in the context and terminology of the humanities, philosophy and social sciences] are [now] raised in the context and terminology of the natural and medical sciences. Much translation and reformulation will be necessary before the rich insights which theology has gained throughout centuries will be of help to biomedical investigators.

EWM explores some theological possibilities along these lines: 'the act of creation is a continuing event; it therefore reveals something not only about the origin, but also about the present history and the *telos*...of this universe'. Again, 'can this concept [of natural law be freed] to become a useful theological clue for an ethics of biomedical research which explores the "laws" and mutations of life, nature and body?'. Yet again, 'can the sinfulness of man, and the cross and resurrection of Jesus Christ, which express and answer the deepest predicament and need of man...become a clue for the understanding of what biomedical research now discovers about the fragility of life, its destruction, and its effective mechanisms of repair?'.

Personal origins

I now turn to *Personal Origins (PO)*. As the subtitle hints, it was produced in the new context created by the setting-up of the Warnock Committee on Human Fertilization and Embryology in 1982. One feature of the working party which deserves some discussion is the discovery of fundamental differences of approach among its members which prevented unanimity. These differences reflected the consequences of the impact of modern science. One approach welcomed the changes and showed a readiness to accept new knowledge.

Another saw humans as in danger of losing any serious sense of the boundaries of natural law and of the purposes for which things are created by God. According to that approach we run the risk of overestimating human abilities and our place in relation to nature. But both approaches seek to arrive at judgments about where the moral boundaries for human action on the natural world lie. *PO* then presents 'certain fundamental principles' which set the terms of the discussion. These principles are 'Creation and Natural Law, Order in Nature, Human Responsibility, Eschatology, The Kingdom of God, Jesus Christ and the Divine Purpose'. It is claimed in the text that the six principles provide the boundaries and, indeed, the avenue for proper Christian reflection and judgment. This is an over-optimistic claim. But the six principles (are they really 'principles'?) provide useful headings.

Of the texts on which I am commenting, PO appears to have been theologically the most ambitious. That it has not succeeded is probably in part the result of trying to weave into the body of the writing a wide variety of standpoints and attitudes and minority views.

Specifically on genetics, the text seeks to rebut the contention that we are 'the prisoners of our genes'. It claims that the genes we are born with do not necessarily determine our life and health. Even in the limited chemical sense we are more than the sum of our genes.

Manipulating Life

Manipulating Life differs from all the other reports under consideration with respect to its expressed view that faith has much to learn from the new science. (Is it in this regard more than a coincidence that a large majority of the working party that produced the report were scientists?) For example: 'how can they [sc. the churches] come to a better understanding of the meaning of faith in the full light of the new knowledge and power that the biological sciences have brought?' 'The creative response to this new knowledge will be one that leads to a deeper understanding of the nature of life and of human responsibility for life.' 'The working out of the relation between faith and this knowledge...we have as yet hardly begun'. This view differs greatly from the initial reactions of some churches, which have been of fear, distrust and even calls for prohibition of these new developments.

ML is also different from the other texts in the degree in which it seeks to identify the background causes of the technological ambiguity. Thus: 'of special concern is the reality that the inequitable distribution of power, influence and knowledge increases the possibility of the misuse of technology'. Indeed, 'many of the social and ethical issues that arise in debates about genetic engineering have less to do with genetics than with broader questions of human good'.

As far as somatic and germ-line therapy are concerned, *ML* reiterates, in its call for caution, what has already been noted in, for example, MIT. But again *ML* takes a further and positive step. It argues that 'nonetheless, changes in genes that avoid the occurrence of disease *are not necessarily made illicit merely because those changes also alter the genetic inheritance of future generations.* By overcoming a deleterious gene in future beings, the beneficial effect of such changes may actually be magnified'. Clearly this addition in *ML,* which is not so remarkable when all is said and done, does not allow us to accuse the text of ethical shallowness.

For, a few paragraphs further on, the theme is taken up again: 'Many ethical and religious traditions endorse some human freedom to modify or transcend nature. But for us to change *substantially* the germ-line DNA is to directly alter the genetic foundations of the human. In what ways do we, by manipulating our genes in other than simple ways, change ourselves to something less than human?'. However, a factor not mentioned in any of the texts treated here may be relevant, namely whether, on the whole, Christian ethicists are more comfortable dealing with individual rather than with collective or social ethics.

Lastly, it is important to take notice of the high degree of methodological openness which characterises *ML,* in particular the relevance of the *unprecedented* nature of many of the problems about genetics. *ML* does not allow that it is feasible, in these circumstances, wholly to depend on 'precedents from the past to provide answers to questions never asked in the past'.

JOSEPH FLETCHER

Fletcher's *Morals and Medicine* is commonly regarded as the first Protestant medical ethics in modern times. However, the text which I am using here is Fletcher's article 'Ethical aspects of genetic controls: designed genetic changes in man'. Fletcher is seemingly a polar opposite to Ramsey. He is a consequentialist, a pragmatist, an empiricist, a situationalist, a personalist, and a (sort of) existentialist. The 'definitive question in the ethical analysis of genetic control' is whether we argue deductively *or* from 'empirical data, variable situations and human values to normative decisions'.

Fletcher, as a consequentialist, argues for genetic engineering from, for example, social needs: 'It is entirely possible, given our present increasing pollution of the human gene-pool through uncontrolled sexual reproduction, that we might have to replicate healthy people to compensate for the spread of genetic diseases and to elevate the plus factors available in ordinary reproduction.' As for the morality of the *means* of genetic research, the only objection would be that fertilization or cloning would result directly and instantly in human beings, or in creatures with nascent or protohuman status. Fletcher denies that embryos are such beings.

Ramsey, Cass and others would argue that the laboratory reproduction of human beings is no longer *human* procreation. For Fletcher the crux is the meaning of *human*. He counters as follows:

Man is a maker and a selecter and a designer, and the more rationally contrived anything is, the more human it is. Any attempt to set up an antinomy between natural and biologic reproduction, on the one hand, and artificial or designed reproduction, on the other, is absurd. The real difference is between accidental or random reproduction and rationally willed or chosen reproduction. In either case it will be biologic -according to the nature of the biologic process'.

Laboratory reproduction is radically human. It is willed, chosen, purposed and controlled-traits which distinguish homo sapiens.

One way of refuting Cass, Ramsey and others of their view is to say that in fact there is, for example, in IVF, very little that is artificial-assisting with the growth and movement of eggs, assisting with the co-location of sperm and egg, and assisting with the journey to the uterus. But Fletcher would want to reduce this (he would say) *high* proportion of biological process if the conception could thereby be the *more* willed, chosen, purposed and controlled. It is because he does not want to sacralize reproductive nature that Fletcher is not disposed to accept that 'in the sequence or progression from aspirin to insulin to artificial kidneys to brain surgery to genetic engineering there is no point at which we can 'change from a clear yes to an absolute no', even though there is a mounting difference in the complexity of the ethical issues posed'.

But the heart of the argument rests upon the definition of the word 'human'. As Richard McCormick observes: 'rational control, therefore, is not the guarantor of human choices but only the condition of their possibility'. It is not clear whether Fletcher has *adequately* considered the 'conflicting interests and values that lie at the heart of ethical problems'.

The texts exhibit different approaches to natural law. The use in PO is virtually unrevised from the Tridentine and postTridentine use. Oliver O'Donovan in *Begotten or Made?:* 'we must cherish nature in this place where we encounter it, we must defer to its immanent laws, and we must plan our activities in cooperation with them' is unrevised. In *EWM,* there is a definition on the road to revision: 'man is aware that he must respond to demands which are proper to his realization of himself as a man, demands written into the very fact that he is a man'. Ryan is clear that natural law does not refer to natural processes and laws found by humans looking in on 'nature' from outside. On the contrary, 'only in man, with his capacity to reflect upon and know himself, can there be any question of natural law, the recognition for himself of how he is to act; and only in his case, with his freedom of action, can there be a responsible following or deviation from the "rules" of his nature...'. So 'it is left for men to discover for themselves the "rules" built into their nature as human beings, to discover how to use themselves if they are to get the best out of themselves and find their proper human fulfilment. And by natural law is not meant these "rules" as built into them...but these "rules" as recognised by men on reflection upon what it is to be human and to find human fulfilment.' This account seems to allow for cultural diversity. Another excellent version for comparison is found in Kelly:

> that mistaken approach to natural law would maintain that by finding out how we function *naturally* as human beings (biologically, sexually, genetically, psychologically etc) we discover how God intends us to live. The view I have been describing does not accept this position. It does not believe that 'meaning' is *written* into 'nature' in this way. It believes that 'nature' has to be *read*-and the very reading is an interactive and creative process in which the human mind plays an indispensable interpretative and creative role.

But though it is clearly possible to draw a distinction, as these quotations do, between a less and a more flexible and anthropocentric definition of natural law, the texts offer little or no indication of what the revised doing of natural law would look like in practice.

The yield of the texts with regard to the purpose and nature of theology is somewhat limited. There are signs of a search for a new theological modality which will be appropriate to genetic and other contemporary issues. The text which is most sensitive to these needs and which tries to say something specific about theological method, namely PO, turns out to be somewhat incoherent theologically. The argument in *EWM* concerning the need for 'translation' of past theological concepts is only convincing to a limited degree. This kind of plea often grossly underestimates the problems inherent in the notion of translation. The need in this respect is actually for a dialectical or two-way hermeneutic-a procedure the difficulty of which greatly exceeds popular notions of what is involved in translation. Another possible direction, also suggested in *EWM*, was that 'for this new kind of theological reflection a team is needed in which specialists of biblical, historical and systematic theology work together with people involved in biomedical and other research'. This proposal recognizes the problems posed by specialism of content. It also recognizes the necessity of exposing the different methodologies to each other. The omission of ethics is strange.

The texts, excepting *MIT* and *ML*, include virtually nothing of the perspectives of the social and political sciences. A terse statement in *ML*, quoted above, tantalizingly points in this direction. 'Of special concern is the reality that the inequitable distribution of power, influence and knowledge increases the possibility of the misuse of technology'. In fact, deductivist theology finds it very difficult to engage with these socio-political dimensions. Of the texts analysed, Fletcher's seems the one which, with modification, promises the most. In particular, the 'situation' to which this 'situationalism' refers has to be taken in a much broader sense than Fletcher allows. But there is nothing about Fletcher's method which precludes the scope of reflection being enhanced by further genres of theological and ethical consideration.

In her useful philosophical reflection on the notion of 'playing God', Ruth Chadwick isolates two types of context. In the second, where the objection to playing God functions as a counsel, it does not seem to be strong enough to rule out particular courses of action. 'It cannot provide us with any definite boundaries which should not be crossed.' I am inclined to say the same of theology and of theological ethics. 'This is [rather] a matter for consequentialist assessment.'

The only tangible ethical dispute raised at any length by the texts concerns the questions about somatic-cell gene therapy and germline therapy. Anderson stated in 1987 that 'essentially all observers have stated that they believe that it would be ethical to insert genetic material into a human being for the sole purpose of medically correcting a severe genetic defect in that patient'. Somatic-cell gene therapy is ethically acceptable 'if carried out under the same strict criteria that cover other new and experimental medical procedures'. These requirements 'simply state that the new treatment should get to the area of disease, correct it, and do more good than harm'.

By contrast, therapy of germ-line cells 'would require a major advance in our present state of knowledge'. The technique has 'a high failure rate', can produce a 'deleterious result', and would have 'limited usefulness'. But, even when the technical problems are ironed out, are there other questions, such as whether 'the transmitted gene itself, or any side effects caused by its presence, will adversely affect the immediate offspring or their descendants'.

All the criteria put forward by Anderson are medical-technical, *except* that concerning 'public awareness and approval'. This reference to public awareness and approval is too weak and too general a way to describe those many, many non-medical areas of ethical perplexity. The latter come to expression very clearly in the Report of the Enquete Commission on *Opportunities and Risks in Genetic Technology*. For example, the Commission took the view that human germ-line therapy 'should be forbidden by law, because, among other things, it might open the way for selective breeding of human beings'. It may be argued that such a consideration is of a universal kind and should apply in all places and at all times. Or, on the other hand, it could be claimed that for the German nation *not* to proscribe in this way would be an immoral symbolic rejection of the memory of the Nazi death camps and of the Hitler doctors. This was the opening consideration of MIT, discussed above.

But there is a counter-argument. So important for the genetic future of (parts of) the human race might germ-line therapy be, that the risk of misuse must be accepted whilst, by the most careful legislation and regulation, everything possible was done to reduce to a minimum that risk. The impression is gained from some of the texts analysed above that there was a sigh of ethical relief that, apparently, there exists a clear-cut choice between somatic and germ-line therapy. But if, as I should contend, the choice is in fact much more complicated, so the range, character and use of the theological and other resources would in turn be all the more complicated. That, however, is to take the ethical challenge *more* seriously, not to undermine it.

I have reserved quite sharp criticism for much of the theological prescription that appears in the texts which I have analysed. This does not necessarily render the texts valueless. It shows, for example, that theology is not a poor kind of science; nor is it an untidy kind of philosophy. On the other hand, theology is not in business to recover and dogmatically to repeat earlier constellations of thought. Theology is better understood as the bringing to conscious form of the varied and sometimes inconsistent lifeexperience of a particular historical-ethical-religious stream or movement, so that this experience can be tested, explored, developed, communicated and shared. The peculiar task is, using the jargon, to combine the experiences of 'historical existence' and 'transcendence' in the area of genetic engineering. It turns out to be a tough assignment. What *that* has to say about 'historical existence' and about 'transcendence' is another story.

Chapter 13

Economic Aspects of Human Genetics

ETHICAL CONSIDERATION

Biotechnology can be defined as the creation of the tools that biologists use to study the genes of animals and plants, both to understand life processes, and to use this knowledge for commercial applications. The Scientific Revolution in seventeenth-century Europe brought together of *scientia* and *tecbne,* knowing and doing so that at this crucial point in western intellectual history the confluence of understanding the world and using that understanding to change it came together to give us the basis of much of our science-based, industrialized world. It is clear that biotechnology is the very visible marriage of *scientia* and *techne,* and because of this we cannot really consider the ethical aspects biotechnology without considering how it is to be used in commerce.

In this paper, I will not discuss biotechnology in terms of the study of the genes of plants, but only in terms of the study of the genes of one small, but extremely interesting group of animals — humans. And I will limit the discussion only to questions of medicine. I use this limited scope to accentuate ethical questions that are of concern not only to professional philosophers and ethicists, but also to scientists and physicians, and to ordinary citizens in society. Because of my background I approach the topic with a completely Eurocentric vision, but

I am aware that the topic requires a more ecumenical approach to be really meaningful.

SCIENCE, TECHNOLOGY AND THE FRAMING OF DISEASE

At the foundation of the association of biotechnology and medicine is the all but universal assumption that the maintenance and improvement of human health is amenable to technological intervention driven by scientific understanding. Science as we know it came late to medicine, and it has had a profound effect not only on how we attempt to maintain health and treat disease, but also on how we frame disease and conceive of health. After the late nineteenth century, when disease was framed in terms of specific causes and for the first time we could think about specific cures, there has been little questioning of the limits of technology in the maintenance and improvement of health. And because of this, until very recently there has been little critical examination of the goals of scientific-medicine. I will argue that this combination of uncritical reliance on technology and lack of questioning of the goals of medicine has led us to potential ethical questions that can no longer be ignored.

In this paper I will address some of the ethical problems in terms of the technology that is coming out of genetics and the Human Genome Project because the great advances in the science of genetics epitomises some potential technological problems we face, and because the Genome Project is the first 'big' science project of modern biology.

Most people would agree with the following dictionary definition of disease: The condition of being (more or less) out of health; illness, sickness. People in all societies and in all periods of history, as far as we can tell, have worried about 'being out of health', and we know that all cultures have variants of their own traditions of maintaining and restoring it. Yet consider these two statement by well-known historians of medicine:

I assert, to begin with, that 'disease' does not exist.

or a bit less shockingly, that

[I]n some ways, disease does not exist until we have agreed that it does, by perceiving, naming and responding to it.

These scholars are telling us that the condition of being 'in or out of health' is not absolute; societies define health and disease, and the definition changes with time, even within the same society. And of course, all societies have selected members who are specialists in how the society deals with health and disease. The European tradition of these specialists comes from the Greeks through Hippocrates and Galen, and is one that for over two millennia had as its foundation the idea that health was the proper balance of humours. Over that very long time, collections of symptoms that were regularly seen were given names and identified as unique diseases, and a hierarchy of healers developed who were adept at bringing the humours into balance by various means.

But there was little consensus through time and among different factions of healers about what had put the humours out of balance in the first place, and what would be necessary to restore their proper balance. While there were specific collections of symptoms that defined a named disease, diseases themselves were not seen as specific entities with specific causes. Any number of things could put the humours out of balance, many of which we had no control over (climate, miasmas, mephitic vapours, etc.) and so any number of things could put them back into their proper alignment. We know that through most of human history, the diseases that people were most preoccupied with — because they were responsible for the 'constant presence of death' — were infectious diseases.

This had a profound effect because now for the first time, around the middle of the nineteenth century, we could think that if there are specific causes of diseases we should look for specific cures. Modern scientific-medicine is medicine based on this assumption; that all disease has a specific cause and that through science we can find specific cures. Disease became the realm of the scientific-physician; it was to be defined by *scientia* and treated by *techne.*

But this relationship which seems so straightforward opens new ambiguities, and takes us to the very heart of the question of what we call disease, how we treat it, and what we expect from medicine. For example, the prevalence in the United States of clinically apparent prostate cancer in men between the ages of 60 and 70 has for many years been considered to be about 1%. But over 40% of men who have no symptoms and are deemed normal according to the standard physical examination of the prostate have been shown to have microscopic changes in their prostate glands that pathologists associate with very early stages of cancer. And with the increased use of new sophisticated imaging technology, tumours too small to be identified by this kind of physical exam can now be seen in even greater numbers.

These are men who exhibit no symptoms of prostate malfunction, or abnormalities during a physical examination; are we to define all of them as having cancer? We had defined the disease by the appearance of clinical symptoms, or the ability of the physician to sense abnormality during the course of a physical examination; are we now to define it as the presence of an image seen by ultrasonography? Should clinicians recommend surgery on the basis of the new tests? If we institute a new definition of the disease the incidence will have gone from 1 to 40% of the population at risk in a few years, does this mean we have an epidemic of prostate cancer?

Remember, the intellectual and emotional impetus for the revolution in framing of disease began when we became convinced that infectious diseases are caused by specific bacteria. We have continued to use this mode of thinking about disease even though we now must deal with diseases that are fundamentally different from infectious diseases. Words like 'epidemic' have entered the language to talk about such diverse topics as the prevalence of breast cancer and urban violence because the idea of a specific cause for being 'out of health' has permeated our thinking. As we enter the twenty-first century we have come to frame as disease, conditions that are the result of our genes. These are not diseases that we 'catch', but diseases that we inherit, and because of our tendency to 'scientificize' and 'medicalize' an increasing number of conditions, we look to the ultimate in specificity, our genes, as causes.

There have been lingering attempts by eugenicists to 'scientifically' categorize behaviour and control the freedom and reproduction of people considered abnormal for many years. Early in 1992 a United States government official suggested that the very serious problem of youth violence, especially in our inner cities, might be addressed by the Public Health Service, the arm of the government responsible for medical research and public health. He suggested that the National Institute of Health examine the genetic component of aggression and violence, and try to find biological markers, especially through brain imaging and genetic testing that would identify those individuals prone to violence. Why even consider a genetic component to youth violence? Because, as the official noted, 78% of violent crimes are committed by 7% of the population, and that population, as every American is repeatedly told, is poor. In the United States the urban poor are preponderantly Black and Hispanic. So on the basis of this social correlation, an apparent genetic association was deduced and a 'violence initiative', similar to the 'genome initiative' that resulted in the Human Genome Project was called for.

Civil rights groups and social scientists immediately claimed that this was a misuse of science; that the social conditions in which those 7% live are the very conditions one would expect to spawn despair and violence among the young. It was pointed out that incarceration rates for black youths was three times that of white youths in 1933, but had risen steadily so that the ratio in 1950 was approximately 4 to 1; in 1960 it was 5 to 1; in 1970 it was 6 to 1; and in 1980 it was 7 to 1. Paradoxically, well into this century one of the traits American whites used as a stereotype of blacks was docility; their acquiescence to the indignities of segregation and other legally sanctioned impediments to full citizenship, and the violence that was perpetrated against them. It has been suggested that if the 'violence initiative' had been initiated in the first decade of this century rather than the last, the genetic correlation would not have been with violence, but rather with meekness. So unless one wanted to argue that evolution was occurring before our very eyes, it seemed only prudent to turn to social factors as the seeds of the problem.

These are just two examples of the kinds of ethical dilemmas we will be creating for ourselves when we to apply twenty-first-century technology to problems we have framed from a nineteenth-century mentality.

Given our growing propensity to look to the authority and technology of science to explain and solve our social problems, and to justify our actions, it is rather predictable that the incredible technical advances that have allowed us to clone and sequence genes has now become the vehicle through which we now want to find specific causes and specific cures. And here is exactly where the danger lies. Technology is only a set of tools, and like all tools, we must suffer the consequences as well as reap the benefits from how we use them. Let me be absolutely clear and explicit; I am not advocating that we do away with scientific-medicine or that technology has no role to play in how we frame and treat disease. My aim is exactly the opposite; I fervently believe that science is one of the great inventions of the human mind and that it is essential for the ordered development of technology as we enter the twentyfirst century. I only urge that we examine the possible ethical ramifications of our actions before we do ourselves great harm. If the unexamined life was not worth living for Plato, the unexamined reliance on technology might make our lives quite different from the ones we hope to live.

And how, you ask, is any of this relevant to the commercial aspects of the Genome Project in particular, and genetics in general? The answer lies in the process by which knowing (science) becomes converted into doing (technology).

We will focus on the pharmaceutical industry, because that is where biotechnology will be converted into products meant to maintain health and cure disease; an industry firmly built on twentieth-century framing of disease, but producing solutions to twenty-first-century problems.

The Visible Hand of the Market

The modern pharmaceutical industry began in Germany at the turn of the century with the discovery of the antibacterial properties of the sulfonamides, and reached its present form during World War II when it was called on to produce large quantities of penicillin. In fact, penicillin, along with insulin and cortisone, formed the industry as we know it today both in terms of the way it operates and perhaps even more important, in the way the public perceives the importance and efficacy of its products. We have come to have very high expectations of the drugs the industry has produced, and the pharmaceutical companies have, until only last year, been the consistently most profitable corporations in the world. Since throughout most of our history the human population has been concerned almost exclusively with infectious diseases, it is understandable that the lay population saw antibiotics as 'miracle drugs' and then transferred this idea to pharmaceuticals in general. Since the late 1940s infectious diseases have, for the most part, been brought under control in the industrialized nations, and the primary focus of biomedical research has been on non-infectious, chronic diseases, and the primary occupation of the pharmaceutical industry has been in developing drugs to treat these diseases. But as I said, we continue to have an infectious disease mentality in a time when we suffer from chronic diseases.

We have of course come to rely on therapeutic drugs, and the pharmaceutical industry is one of the best examples of society turning over the control and production of an important need to the free market economy. We function on the assumption that factors that affect our health form an industry, and that the forces of the market will ensure that the industry will meet the needs of society. This idea has been a basic tenet of the market economy (Adam Smith's 'invisible hand of the market'), and the pharmaceutical industry worldwide had been assumed to be dramatic proof that the forces of the free market are sufficient to serve the needs of the society.

The recent revelations that the industry spends more money on promoting than developing drugs, and that over half of those drugs developed in the last decade are 'me too' drugs that do not add any therapeutic advantage, has raised questions about just how 'invisible' the hand of the market really is. The assumption that consumers will pay for what they need and want, and that any product that is either un-needed or un-wanted will fall the test of the market has been called into question by this great expenditure of funds on promoting drugs. In one issue of the *New York Times*, the news section reported the fact that more money is spent on advertising (i.e. creating a market) than on research, the financial section stated that the fall in price of the stock of a well regarded biotech company was due to the fact that the market for one of their life-saving drugs (designed of course to treat a specific condition) was not as large as had been predicted and was an indication that the company would have to begin to advertise to create a greater market for their drug (i.e. to treat other conditions).

So in a free market economy in which it is almost an article of faith that the choice of products by the pharmaceutical industry will be determined by the *invisible* hand of the market, it is important to understand how priorities are set and decisions are made about which drugs to develop, test, and market. We must realise that the hand of the market is in fact, very *visible* and look to how it will treat the explosion in genetic information science is giving us.

Established pharmaceutical firms and new start-up biotech companies each face variants of the same fundamental question; will there be a sufficient *return on investment* to justify the development of a product? A start-up biotech firm must attract investors with the promise that there is a large enough market for the potential products to provide an attractive return on their investment, and when the time comes to sell their stock to the public, these companies must also be in a position to assure the larger investing community that there will be sufficient profit to justify investment. Large, established firms also allocate their internal funds based on the potential return on investment; when they decide which of many potential products to develop to keep a flow of products in their 'pipe-line', those with the greatest potential return have a better chance of being developed. The size of the market is therefore crucial in determining if start-ups and established firms will enter into a market. It is always a surprise to those who are not in the industry to realise that market sizes are almost always given in terms of dollar values, not cost of manufacturing and development, incidence of the disease, social needs, etc. (although any of these may be factors in determining the dollar value).

If the dollar value of the market is not large enough to assure a market that will give the desired return on investment, the companies have three options. They can decline to enter into the market, they can raise the unit price of the drug, or they can create an increased need for the drug. But if they choose the third option, there must be enough financial resources in the market they are entering to pay for the product to give the return on investment. Payment is increasingly assured by governments and by private insurers, and governments and insurers are increasingly under pressure from single-disease oriented pressure groups. In the United States, private foundations devoted to calling attention to a particular condition like diabetes, arthritis, or cystic fibrosis have played an increasing role in exerting pressure on insurers and government (a topic that deserves much fuller exploration).

Let me give one example, that of end stage renal disease (ESRD), once known as kidney failure, and before that as Bright's disease. Payment for ESRD treatment by dialysis has for two decades been mandated by federal law, so the diagnosis of ESRD has ceased to be only a medical classification of interest to patients and their doctors, but has become a guarantee of financed treatment. Manufacturers of dialysis machines and reagents, surgeons and hospitals involved in kidney transplantation, and a whole economic apparatus has grown up around this legislative decision.

So needs (i.e. markets) can be created in a variety of ways, all of which make the idea of the 'invisible hand' of the market unrealistic. Assuming that there are not inordinate costs projected in the development and testing of a new drug, and that patent protection is adequate for a period of time that will ensure the desired return on investment, what is mythologized as the 'invisible hand' of the market is really aided by the perceptible hand of the manufacturers and special interest groups. The question that we must then ask is, what is there to commercialize from the Genome Project that is relevant to all of this? How can sequencing the entire human genome lead to any products that even those most adept at creating markets can work with? The answer is genetic testing.

Genetic Testing

In the practice of modern scientific-medicine, the decision to treat a patient with a drug is made by the physician, almost always based on one or more laboratory tests.

In fifteen years, we will probably be able to apply a single multiplex test to fetuses in utero, babies at birth, or in many cases, parental carriers, a test that will detect somewhere between 100 and 1,000 of the most common genetic diseases, disease predispositions, and genetic risk factors for environmental insults, drug dose responsiveness, and the like. We will be able to do this extensive fingerprint for any individual, but we will, at least initially, be unable to offer any help based on this information.

So according to this molecular geneticist who is close to the Project, by the early years of the twenty-first century we will have simple tests that will detect the most common genetic diseases, predisposition to those diseases, and the genetic risk factors for environmental insults and drug dose responsiveness. If the twentieth century saw the flowering of biological research, surely the fruits are to be harvested in the twenty-first. What more could science promise than the ability to predict both disease *and* predisposition? But remember, these are promises of solutions for problems that have been framed with an infectious disease viewpoint.

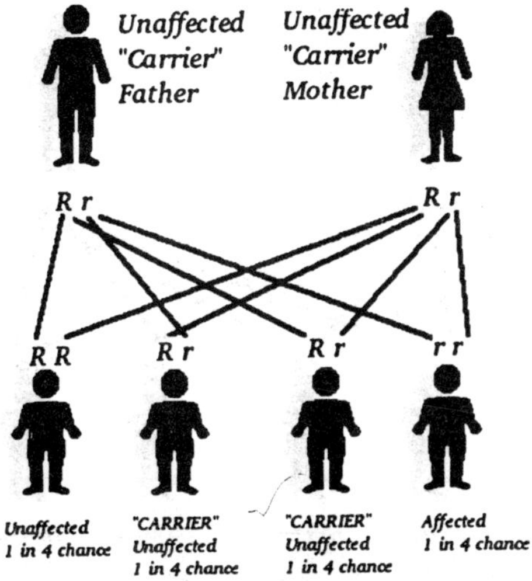

Fig. Genetic Testing Genetic Markers

To fully understand the implications of this promise, we must understand two things about our current understanding of genetic diseases. The first is that the diseases for which the tests described above will have predictive value and are caused by a *single* gene defect. There are genetic associations of a great many diseases, but in most of these cases there are *multiple* genes involved, and we have virtually no idea which of them is important and predictive.

The second point is that most diseases due to a single gene defect are relatively uncommon; the most common having incidences of around one in 2,000 births. But the most striking thing about them is not so much the number of cases per 1,000 births for any particular trait, but the sheer number of different genetic diseases. The probability of a particular individual being affected by any one of them may be small, but the large number of known, single-gene genetic diseases strikes fear into the heart of any prospective parent. The promise of a test to determine if the parents are carriers of a recessive gene or if the fetus has a rare disease is the fulfillment of the promise of the predictive power of medicine to prevent disease. Some of the singlegene diseases that are currently available for test include:

Fragile X syndrome sickle-cell anemia Duchenne muscular dystrophy cystic fibrosis Huntington's disease hemophilia Gaucher and Tay-Sachs

There are less than 1 million people in the United States who suffer from any one of these diseases. This is roughly 2 or 3 per 1,000 of the population.

Now what about the promise that we will be able, with a single test, to determine *predispositions* to disease and genetic risk factors of environmental insults? One reputable source of such potential future tests lists the following susceptibilities to which we can look forward to being screened:

Hypertension dyslexia atherosclerosis cancer manic-depressive disease schizophrenia type 1 diabetes familial Alzheimer's

There are 90 million people in the United States who suffer from one of these eight diseases. This is roughly 500 people per 1,000 population, or one person in every two.

To understand the ethical significance of these numbers, it is necessary to know that in these diseases for which there are 'potential future tests', there is a suspected genetic element (or predisposition) but none of these conditions are caused by a single gene. In other words, if the tests were to be applied to these conditions before it is known which of the genes involved have predictive value for the condition, the predictive value of the tests themselves would be questionable at best.

Now it does not take a knowledge of Aristotle's *Ethics* to realise that to use a test for the genetic 'predisposition' to a disease that has little predictive value because the genetic basis of the predisposition involves many genes, most of which are as yet unknown, is fraught with problems. What is a law firm to make of test results that show that a fifty-year old attorney at the peak of her powers has a 'genetic predisposition' to Alzheimer's disease? Or an airline of the fact that a young pilot in otherwise perfect health has a 'genetic predisposition' to schizophrenia? But if the tests have such low predictive value, why would anyone worry that they will be developed?

Will not the forces of the market guarantee that a product with questionable value will not survive? To approach that question, we must return to the factors that will affect the people who will do the developing, manufacturing, marketing, selling, and will be making the profits to return to their investors. I will not insult this audience by suggesting that the free market is an evil cabal that will foist anything on an infinitely malleable public; industry and society often act in concert to create the need for a product. My point is that we must be aware of the factors in society that would act as the 'visible hand' because to a large extent we will be helping to create our own ethical problems.

The Role of the Law in Creating a Market

There are many ways to market the need for massive genetic screening, but one of the most probable is through changes in the law.

In the UK under the Congenital Disabilities Act of 1976 legal action can be brought against a person whose breach of duty to parents results in a child being born disabled, abnormal, or unhealthy. In both the UK and the USA there has been an escalation of litigation concerning genetic disease. Most cases concern physician errors of omission, and it behooves all doctors who provide genetic advice to ensure the validity and up-to-datedness of that information. Thus failure (whether from ignorance or religious objections) to give correct advice, or to refer the consultand to someone who will advise, on the risk of fetal abnormality in a future pregnancy and the possibility of prenatal diagnosis, may constitute medical negligence...*Further, in the USA a physician may now be considered negligent for failing to take a history which includes relevant ethnic and family information.*

If one aspect of the practice of 'defensive medicine' is to make sure that all patients have undergone a full screen for possible genetic defects, including multifactorial disorders, one need not be a trained economist to see that the market is enormous, and that the economic pressures on corporations to create the atmosphere in which their products can prevent legal actions will be great.

Prospects for the Future

The number of diseases due to mutations at a single gene is not enough to create a market of sufficient size to give the kind of return on investment for gene testing the free market demands. In the US, the total annual market would be around $100 million (at $100/test), the minimum dollar amount for a single blockbuster pharmaceutical. The close to 100 million tests available for conditions in which more than one gene is involved is what the market industry is interested in ($10 billion at $100/test). But for most of the conditions in that list it is not clear that we are close to identifying which of the multiple genes (if any) are predictive, and recent research does not hold out much hope that they soon will be.

The results from 'gene knock out' experiments in mice are showing us that when tested *in vivo,* even the genes whose functions we think are well understood, almost invariably give surprising results causing me to have reservations about the rate at which we will understand the function of multigenic phenomena. If my pessimism is justified, then we must put our faith in the fact that the forces of the market will be able to resist the temptation to wait until science clearly demonstrates that a particular set of genes is unequivocally associated with conditions before marketing a test.

The problem can be summarized as follows: *large scale* genetic testing of the population is the only way investors can get their return on their money, and society must guard against genes being tested that do not actually predict a tendency or propensity for the condition. The rate of converting science into machines is rapid, that of converting it into drugs is slow;

the technology for large scale screening exists and short of legislation, I see no barrier to the visible hand of the market coming into play rapidly. I also see no reason for optimism that advances in genetics will in the near term be able to *reliably* predict alcoholism, schizophrenia or multiple sclerosis by gene testing. I have great fears that whatever genes are available will be screened and large numbers of people will be placed in financial and emotional jeopardy as a result.

The rate of research aimed at understanding the genetic and molecular basis of normal and abnormal metabolic functions is proceeding at a rapid rate and is providing scientists with new insights and surprises. It is already clear that the vast majority of genes that contribute to our development and affect our responses to environmental agents or influence our behaviour act in complex and integrated manners. This research is already revealing redundancies in gene function and properties of emergent systems that will take decades for scientists to fully understand. The Human Genome Project as a major contributor to the interest of the public in gene therapy is aware of its responsibility in educating the public against premature expectations of as yet unproven therapies that have already come from this research, and ethical dangers in the testing for genetic traits that will follow, and has set aside a significant amount of money to explore the ethical and social questions raised by the project.

Historically, commercial forces have created the markets for new technology and we see that the same is true for the commercial uses of the Genome Project and all genetic research. In the present case, however, the commercial forces have been combined with the unbridled enthusiasm of scientists, patient groups lobbying for therapies for individual disease, and a generally uncritical press. Unfortunately, there is no mechanism for the public to make informed decisions if this is the direction it wishes the expenditures in health care to follow. Legislation may be needed for the short term, but society must begin to re-frame disease in light of modern scientific discoveries so that it can have more realistic goals from medicine and the technology that will flow from science.

